Alfred Waddington

Overland Route through British North America

Or, The shortest and speediest Road to the East

Alfred Waddington

Overland Route through British North America
Or, The shortest and speediest Road to the East

ISBN/EAN: 9783337148300

Printed in Europe, USA, Canada, Australia, Japan

Cover: Foto ©Lupo / pixelio.de

More available books at **www.hansebooks.com**

OVERLAND ROUTE

THROUGH

BRITISH NORTH AMERICA;

OR,

THE SHORTEST AND SPEEDIEST ROAD TO THE EAST.

WITH A COLOURED MAP.

BY

ALFRED WADDINGTON.

"Once lost, never regained."

LONDON:
LONGMANS, GREEN, READER, AND DYER.

1868.

Price One Shilling.

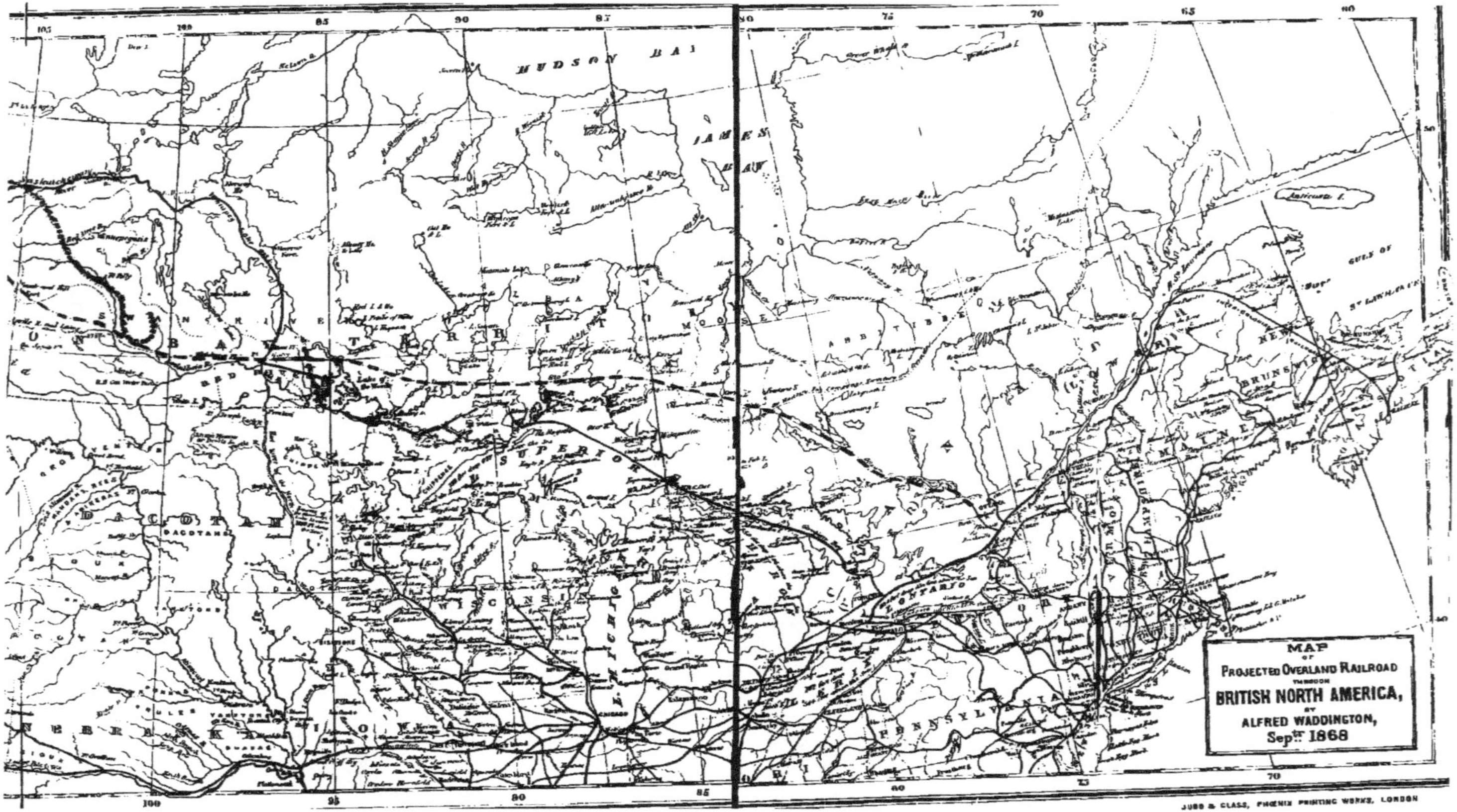

JUBB & CLASS, PHŒNIX PRINTING WORKS, LONDON

OVERLAND ROUTE

THROUGH

BRITISH NORTH AMERICA;

OR,

THE SHORTEST AND SPEEDIEST ROAD TO THE EAST.

WITH A COLOURED MAP.

BY

ALFRED WADDINGTON.

"Once lost, never regained."

LONDON:
LONGMANS, GREEN, READER, AND DYER.

1868.

JUDD & GLASS,
PRINTERS
LONDON

PREFACE.

A portion of the following pages was read at the late meeting of the British Association for the Advancement of Science, in Norwich. I also read a paper in connexion with the proposed route at the Royal Geographical Society, shortly after my arrival in England; and in both cases the subject excited considerable attention, and was reproduced in most of the leading papers. I have been led to embody the whole in the present *brochure* since my return from Norwich, at the request of several well-known gentlemen, who feel and understand the importance of the question at issue; for it is by publicity alone that our Government and the nation can be aroused to its importance. The subject is a most serious one, and I recommend it to the earnest consideration of our bankers and merchants, to that of the different chambers of commerce, and of the mercantile community generally.

ALFRED WADDINGTON.

Tavistock Hotel, Covent Garden,
September 17*th*, 1868.

OVERLAND ROUTE THROUGH BRITISH NORTH AMERICA;

OR,

THE SHORTEST AND SPEEDIEST ROAD TO THE EAST.

"Once lost, never regained."

SUCH is the motto with which I head these pages, because it embodies a great truth, and an ominous one, as regards the subject about to be discussed, namely, "The Shortest and Speediest Road to the East." The unprepared reader may feel surprised, but if he take the trouble to go through these pages, he will soon admit the correctness of its application. The truth is, that England is, commercially speaking, on the brink of a precipice without being aware of it. Wrapped up in her own prosperity, she is apparently ignorant that a trans-continental railroad is rapidly progressing through the United States, for the professed purpose of transferring the trade of the Old to the New World, and that ere long it will be completed. Or, if not entirely unaware of the fact, the few persons who have turned their attention to it are either heedless as to its general importance, or else foolishly incredulous, and therefore indifferent as to the results. And yet England has in her hands the means of rivalling that high road, as we shall presently show, by one still more direct through British North America, and of thus averting the impending danger. But this is not generally known; or rather (as is almost always the case when a subject that is not over favourably looked upon, has been but imperfectly studied, and that from a distant point of view), those who are more or less aware of it are only acquainted with the objections that have been made to its practicability. And these have been so often repeated, that they are almost taken for truth. In the meanwhile the Americans are actively advancing towards the attainment of their object, and if England neglects this opportunity, she will awake, when too late to recover them, to regret the loss of her trade with the East and her commercial supremacy.

But before proceeding further, it may be well to give a short description of the way in which this great American enterprise, the most magnificent perhaps of modern times, is progressing, so that the reader may better realize the extent and imminence of the danger with which we are threatened.

CENTRAL PACIFIC RAILROAD.

Although the Railway in progress from New York, or more correctly speaking, from the Missouri to the Pacific, is in many places accompanied by great engineering difficulties, and passes over a vast tract of country unfit for settlement, yet, in spite of these drawbacks, it is advancing with the most astonishing rapidity; owing to the liberal assistance afforded by the Federal Government to the two companies (the Union Pacific, and the Central Pacific), who have the concession, and to the profound public conviction of the immense results to be obtained. On the Eastern side, from Omaha, on the Missouri, to Cheyenne, at the foot of the Black Hills, 517 miles, of which only 40 were completed on the 9th of May, 1866, had been laid down in December last. These 517 miles pass over the rich plain and valley of the Platte river, where rising villages, and towns containing some of them hotels and all the proofs of active business and prosperity, already show the progress of the line; and, amongst other considerable works, include a bridge 3,000 feet long over the North Platte, and a noble Howe Truss bridge, 1,500 feet long, at Loup Fork. Two thousand graders, accommodated in cars, 80 feet long, fitted up with berths, or as dining halls, kitchens, store-rooms, etc., which follow the terminus, and are pushed on as it advances; and 1,500 wood-choppers and tie-getters in the woods formed the advanced guard at this end, besides all those employed in laying down the line. The rolling stock employed consisted of 793 freight cars, 20 passenger and mail cars, and 53 locomotives; and on an average over one mile and one-third of track was laid per working day—a speed unparalleled in railroad building. After the season had closed, the rock cuttings and gradings were continued during the winter to the summit of the Black Hills (31½ miles further on, and 8,262 feet above the sea), and thence over the Laramie Plains to Bridge's Pass (7,434 feet), in the Rocky Mountains, the difficulties over which extend about 150 miles. At this point the works were resumed early this spring, and are at this moment proceeding with still greater rapidity than ever; whilst further west, the portion in the Desert, from Echo Pass to Salt Lake, has been lately contracted for by the well-known Mormon Elder, Brigham Young.

On the Pacific coast (where 10,000 Chinese were employed by the Central Pacific Company at one dollar per day) the road had reached and crossed over the Sierra Nevada in December last, at a height of 7,042 feet, after surmounting the greatest difficulties perhaps on the whole line. Some idea of these may be formed from the fact, that besides fifteen tunnels and numerous heavy mountain side-cuttings, many miles of road have had to be covered in with sloping roofs, made of the strongest timbers (an entirely new idea in railway construction), in order to protect these places from the danger of avalanches, which might otherwise bury the trains and sweep everything before them. About 105 miles had been completed at this end last winter, from Sacramento to Mud Lake, on the eastern and more abrupt slope of the mountains; 64 miles more had been added this year at the end of

June, and it was confidently expected that the whole grand line to the Pacific would be completed in 1870, so as to connect with the splendid steamers already subsidized by the American Government, and running between San Francisco, Japan, China, and the Sandwich Islands. So confident, indeed, are the Union Pacific Company that this line will monopolise the commerce of the East, that closed cars are already built, as if the service were secure, and the closed pouches only want labelling for "China through," "India Official," "Sandwich Islands," "Alaska," "Japan," &c. (See the Report of the Company, New York, October 25th, 1867). Making all reasonable deductions from these exaggerations, no one yonder expresses a doubt as to the success; and in San Francisco, such was the influence of the same conviction on the merchants and others, and their confidence in the results, that when the writer was there a few months ago, palaces were literally rising up as if by magic. But what must be the feelings of every Englishman, when trying to calculate the consequences of such a commercial revolution! One which, unless counteracted, will at the very onset throw the Chinese trade, and that of Japan, into the hands of the Americans. The precious metals, the transmission of which to the Oriental ports has been hitherto by way of London, will in future be sent at half cost by this more speedy and direct route; thus making New York and San Francisco, instead of London, the financial and banking centres of the trade of the world. The business of all those of our merchants who are at present engaged in direct trade with those countries, will be disturbed—if it be not wrested from them; our communications with New Zealand and the Australian Colonies displaced and thrown into foreign hands, and the general inroad into our commerce with the East will sound the first knell of England's decline.

Having thus cast a hasty glance at what the Americans are doing, let us now examine *seriatim*, and more attentively, the difficulties which are supposed to render the construction of a rival railroad through British America problematical, if not almost chimerical.

OBJECTIONS TO A RAILROAD THROUGH BRITISH TERRITORY.

These are numerous enough, if true, and may be classed as follows:—

1st.—The supposed geographical difficulties to the north and west of Lake Superior, and those much more real through the Rocky Mountains and British Columbia.

2nd.—The supposed severity of climate, and general unfitness of the country to be traversed for settlement.

3rd.—The greater distance across the Continent to the North, as compared with the South.

4th.—The difficulty, if not impossibility, of constructing a railroad through a wild, unsettled country.

5th.—The opposing rights of the Hudson's Bay Company.

6th.—The possibility of difficulties at some future day with the United States, combined with the existence of a sparse population and a long line of defenceless frontier; and consequently, the little confidence placed by English capitalists in anything appertaining to Canada.

7th.—The enormous expense of the undertaking, supposing it practicable.

8th.—The cost of railroad transit across the Continent, and consequently the small amount of traffic to be expected.

9th.—And last, not least, the anti-colonial theories of the day, and the growing dislike to spend money on our foreign possessions.

This is, however, not a Colonial, but a commercial and imperial question which concerns the whole nation. That unimaginative Englishmen should not fight for an idea, may so far show their good sense; but that they should be so indifferent and shortsighted as to give up that commercial supremacy which has been so slowly and dearly acquired, and on which the greatness of this country depends, without a struggle to retain it (and that not a struggle by arms, so as to expose the country to the uncertain risks of war, but one of commercial rivalry); or merely in order to avoid the temporary burden of an expenditure, the wisest England could ever make, is a thing not to be understood. We must, therefore, look into the matter more closely as the importance of the case requires, and weigh the above objections one by one, in order to show how groundless they are; beginning with the geographical difficulties, as an indispensable preliminary to the others.

GEOGRAPHICAL DIFFICULTIES.

It has hitherto been generally believed, for want of more ample information, that the country north of Lake Superior was broken and barren in the extreme; thus rendering it unfit for settlement, and consequently to serve for an Overland communication with the west. So that the only feasible road to connect Canada with the North West Territory and the Pacific, must unavoidably be through the State of Minnesota.

Such a conclusion can only have been founded upon the forbidding aspect of the mountains which form the northern shores of Lakes Superior and Huron; and which, as seen by travellers from the water, with their bold naked sides and peaks, treeless and bare of vegetation, present, it is true, a scene of thorough desolation. But the explorations which were made last year in that direction by the Canadian Government (the results of which were kindly communicated to the writer by Mr. Russell, Crown land agent in Ottawa), prove that this apparently formidable range of mountains has no breadth, and is as circumscribed in a northerly direction as its southern flanks are precipitous. So much so, that at one point the watershed towards Hudson's Bay comes within eight miles of Lake Superior; whilst to the north lies a vast level country of clayey formation, extending with little interruption to Hudson's Bay. Good crops of wheat are raised at New Brunswick House, on Moose river, in lat. 49·35, and as the level tract of country south of this is (with the exception of some portions north of the Montreal river, which are poor and sandy), of much the same quality as that of the Ottawa country, it may be safely inferred that the whole country is fit for settlement.

But the facilities for a railroad are still more remarkable. From Ottawa to the mouth of the Montreal river, 280 miles, the country, which is well known, presents no serious obstacle. The watershed at the north angle of the Montreal river, 100 miles further on, in lat. 48·6; long. 81·20, and the highest point between Ottawa and (probably) Nipegon river, is only 830 feet above the sea. The ground explored here for 105 miles due west, and to within a distance of 280 miles of the river Nipegon in long. 88·25, was found to be "most favourable," and the surveys made in the latter neighbourhood and extending twenty miles back from Lake Superior, show the country to be "still more even."

It has even been suggested, that a railroad might be carried in a direct line from Quebec to Nipegon river, along the watershed between the St. Lawrence and Hudson's Bay; thus avoiding 300 miles of frontier line by the existing railroad along the St. Lawrence, and shortening the total distance across the Continent some 120 miles. But there has been no survey of this portion of the country; the length of road to be built would be increased 250 miles, entailing an extra expense of two millions and a half sterling; and besides this very serious objection, and those arising from the absence of population, a more northern latitude, and greater elevation, the Port of Quebec is closed during the winter, thus presenting no advantage over that of Montreal.

Further west, a tract of country between the Nipegon river and Sturgeon Lake, where the rock-formation is Lawrentian, and another nearer the Winipeg river, north of the Lake of the Woods, with numerous dome-shaped hills composed of intrusive granite and syenite, varying from 150 to 200 feet high, might offer some difficulties, besides a considerable amount of sterile ground. But those immediately between Thunder Bay on Lake Superior (where a silver mine of surpassing richness has lately been discovered), and the Lake of the Woods, are not by any means what has been said, or what is still very generally believed. The whole country in this direction was carefully explored in 1858-9 at the expense of the Canadian Government, and a line of communication to Fort Garry laid down by Mr. Dawson, the well-known engineer; by Dog Lake, Savanne river, the Lake of a thousand lakes, the river Seine, Rainy Lake and River, and the Lake of the Woods; inall 499 miles, of which 308 are navigable by steamers. The opening of this line and building of a dam at Dog Lake, were commenced last year, but suspended soon after the installation of the new Dominion. It would, by taking advantage of the lakes and rivers, cost (according to the plan which might be followed), from £50,000 to £80,000, of which the Red River Settlement would contribute a part; and Mr. Dawson calculates, that it would reduce the cost of conveying goods to Fort Garry, to less than 40 dols. per ton from Lake Superior, as against 100 dols. from York Factory, and 90 dols. from St. Paul, Minnesota; besides a saving of 30 per cent. on the value of the goods, by buying them in Canada instead of in the United States. These prices would again be eventually reduced by the construction of the Huron and Ontario canal, for which a company has been organised and authorised by Act of the Canadian Parliament.

These geographical facts, some of which are laid before the public for the first time, settle the question as to the supposed preference to be given for any future road to a line through Minnesota (where the Red River settlement at present gets its supplies); and which, instead of being the "true and only practicable route from the North Atlantic to the Pacific," as some parties have maintained, would in all respects be by far the most roundabout. A railroad from Ottawa to Fort Garry, passing north of Lake Superior, as above described, would not only form one single straight line in the direction of the Yellow Head Pass through the Rocky Mountains, but would pass entirely through British territory, and at a suitable distance from the frontier.

Further west, the prairie country and great plain of the Saskatchewan (the best access to which is thus shown to be in our own hands), extends from the Lake of the Woods to the foot of the Rocky Mountains; presenting one thousand miles of the easiest ground in the world for the construction of a railroad, and of the most admirably suited in point of climate and fertility for settlement. Unlike the arid American desert, inhabited by hostile Indians, the proposed line would pass here over one of the richest, most beautiful, and fertile regions in the world, containing more than 60,000 square miles, or over forty millions of acres, clear, and ready for the plough, lying directly between the Canadian Dominion and British Columbia, and possessing every qualification for agricultural purposes. A line of communication, where prairies covered with luxurious grasses are mingled with stretches of woodland, and watered by numerous lakes and streams, and which would soon be followed up and fed by an agricultural population from one extremity to the other. Indeed for settlement there remains nothing of the kind to be compared with it, either in the United States or British North America. (See the Report to the New York Chamber of Commerce, Appendix A.)

Beyond this beautiful plain, and further west, we come to the Rocky Mountains, which form the limit of British Columbia, and to those which compose the greater part of the interior of that colony. But here the difficulties to be surmounted are far more serious than any we have yet had to deal with, and compared with them, those around Lake Superior are child's play. Indeed, for some years it was a matter of discussion, and even of doubt, whether any available communication for a railroad could be found through them. It was only by a series of lengthened and expensive explorations, that a practicable road through the Cascade, or Coast Range, was at last discovered by the writer, so as to communicate by the valley of the Upper Fraser with the Leather or Yellow Head Pass, in lat. 52·54, through the Rocky Mountains. After which, a careful investigation of the explorations made by Palliser, Hector, Blakiston, Sullivan, and others, of the different passes to the south, and nearer the Boundary line, having convinced him of their general impracticability; and the impossibility of connecting them with any good harbour on the Pacific, having been, moreover, well established; he came to the conclusion, that the northern route by the Yellow Head Pass, and then over the Chilcoaten Plain to Bute Inlet, was by far the best, and, indeed, the only feasible one for a railroad to the Pacific. His reasons for such an important decision may be very properly inserted here, and summed up as follows:

1.—The arid nature of the country traversed by the South Saskatchewan, the greater part of which is unfit for settlement, its proximity to the Boundary Line, and the hostile disposition of the Indians.

2.—The much greater altitude of the Passes, the sharpness of the grades and curves, and the greater amount of snow.

3.—The circuitous course the route would be obliged to follow through the western portion of the Rocky Mountains, after having crossed the main crest or watershed; amounting to nearly 250 miles of most expensive if not impossible railroad.

4.—The enormous expense, if not impossibility, of carrying a railroad in this latitude through the Cascade, or Coast Range, and down the Fraser to New Westminster.

5.—The utter worthlessness of the greater part of the mountainous country thus traversed, amounting to at least 450 miles out of the 600 by Howse Pass.

6.—The difficulties of access to the port of New Westminster, which render it totally unfit for the terminus of an overland railroad. (See foot-note to extract B, Appendix, page 32.)

PER CONTRA.

1.—The well-known fertility of the whole country drained by the North Saskatchewan, and commonly called the Fertile Belt.

2.—The greater navigability of the North Branch, and the presence of large seams of coal on several points.

3.—The natural connection of both with the road by Jasper's House, and the Yellow Head Pass, and the facility of the latter, which requires no tunnel. This pass, or rather valley, presents a natural break through the Rocky Mountains; its greatest altitude is only 3,760 feet above the sea; the Indians cross over it in winter, nor does the snow render it impassable at any time.

4.—The ready and easy communication offered for 280 miles by the Upper Fraser and its valley, through a comparatively open and fertile tract of country.

5.—The opening up of the gold mines in and around Cariboo, which at present can only be reached by 380 miles of wearisome, mountainous waggon road; so that only the very richest claims have been hitherto worked.

6.—The opening up of the Chilcoaten Plain, the only one of any extent in British Columbia, and which contains millions of acres fit for settlement.

7.—The facilities offered by the Bute Inlet Valley, presenting a level break, 84 miles long, through the Cascade range, and the *only one* for constructing a railroad to the salt-water.

8.—The superiority of the harbour at the head of the Inlet, its proximity to the coal mines at Nanaimo, and its easy and safe connection with Victoria, Vancouver Island, and the ocean.

The great difficulties which exist on this portion of route through British Columbia, and the way in which the writer succeeded in surmounting them, were explained for the first time in a paper read by him at the Royal Geographical Society, in London, March 9th last, from which, as they form an important part of the present subject, an abstract has been made, which the reader can consult, in the Appendix (B), so as to satisfy himself of the result.

It has thus been conclusively shown, that the geographical difficulties which have been so much talked of, through British America, either do not exist or can be avoided; so that there no longer remains a doubt, as to the facility of constructing a railroad across the Continent in almost a straight line from Ottawa to the Yellow Head Pass, and thence to the Pacific. Indeed the general facilities for that purpose are as great through British territory, as the difficulties on the American line are considerable. And here it may be observed, that whilst San Francisco possesses no coal for steamboat purposes, the termini of the English line, both at Halifax and Bute Inlet, would be abundantly provided with it. It is hardly necessary to add, that from Ottawa the proposed line would connect by the existing railroads with Montreal in summer, and when that port is closed, by the intercolonial railroad with Halifax on the Atlantic, or perhaps Shippegan in the Gulf of St. Lawrence. The most direct line however to the Atlantic in winter would be by Portland.

SEVERITY OF CLIMATE.

This has also been exaggerated, till at last one would suppose that Canada was fairly locked up and buried in snow and ice for seven or eight months in the year. But the Lechine Canal, near Montreal, remains open on an average for 226 days, or 7½ months, whilst the St. Lawrence at Montreal is navigable still longer. As far, however, as the present subject is concerned, the objection mainly embraces the running of trains in winter, and the fitness of the country for settlement. Now the amount of snow is the only serious obstacle to the running of trains in winter; and because the winters in any country are severe, the fall of snow is not necessarily greater. In Canada the greatest depth of snow is to be found in the maritime provinces east of Quebec, where it is occasionally but rarely known to lie from four to five feet deep; but south and west of Quebec this is no longer the case. As a general rule, the snow in Canada is easily removed by the snow-ploughs, which are used both there and in the Eastern States, and the trains run regularly all winter, with the exception of an occasional snow storm. But as we get farther into the interior, the thickness of snow continues to diminish with the decrease of atmospheric moisture, till in the plain of the Saskatchewan it does not pack over fourteen inches thick in winter, and then evaporates quickly; and even in the Yellow Head Pass in the Rocky Mountains, it barely attains from two to three feet. In addition to these facts, the isothermal lines, which run in a W.N.W. curve across the Continent, show an increase in the mean temperature on the Pacific coast equal to fully 11 deg. of latitude as compared with the Atlantic; whilst the range of the thermometer becomes less, and the winter and summer temperatures more equable. Thus, the mean annual temperature at Cumberland House in lat. 54, long. 101·40, is only one degree lower than that of Toronto, 10 deg. more to the south, but also 42 deg. more to the east; and in Victoria, Vancouver Island, where snow rarely falls, and the arbutus grows in the open air to the size of a tree, the climate closely resembles that of Nantes or La Rochelle in France. In short, if the trains run all winter in Canada, they could do it *a fortiori* across the western portion of the Continent.

As to the general fitness of the country for settlement, that has already been shown as regards the great plain and "clayey level," extending together for 765 miles from Ottawa to Nipigon river. Beyond this, there is an interval of 285 miles, between the Nipigon and Winipeg rivers, a small portion of which, as already explained, is composed of silurian rocks, and comparatively sterile. But although the cultivable areas are limited, where they do occur, the soil is rich, and the country is intersected by many fertile spots and hollows sufficiently extensive for farms. Further west, the beauty of the Fertile Belt, which stretches in a north-westerly direction for one thousand miles, has at last been recognised, and is now becoming world-renowned; it has truly been named a "Paradise of fertility," and its soil and climate require no further comment. Indeed, its climate is more suitable to the emigrant from Northern Europe than that south of the Missouri, where summer droughts are common, together with excessive winter colds and winter snows.

In British Columbia there exists a large tract of fine country along the Upper Fraser; and farther west the proposed line traverses the great Chilcoaten or Central plain of the colony; a garden of itself, full of agricultural and pastoral wealth, and containing over twenty millions of acres, the two-thirds of which are fit for cultivation. When we compare this succession of fertile lands with the sterile regions of the American desert (though traversed by the Central Pacific Railroad in one of its narrowest and least arid portions) and the facilities of the British line over the American in an engineering point of view, we may feel ashamed to think, that we have made so little use of the superior advantages at our disposal, and that the Americans, under far greater obstacles, have got so far ahead of us. (See Appendix C for a description of this desert.)

COMPARATIVE DISTANCES ACROSS THE CONTINENT.

The following table shows that the distance across the Continent from Montreal would be less than that from New York to San Francisco :—

			Miles.	
Distance from New York, by Chicago, to Omaha, on the Missouri			1531	
From Omaha to San Francisco (of which the two-thirds comparatively sterile), about			1830	3361
From Montreal to Prescott	133			
From Prescott to Ottawa	54	187		
But these two railroads form an acute angle at Prescott, which would be avoided by building a railroad in a direct line from Montreal up the Valley of the Ottawa, at a greater interval from the frontier, and shortening the distance		72	115	
From Ottawa to Bute Inlet, as shown further on, about			2885	3000
Difference in favour of Montreal				361

In winter the increased distance to Shippegan or Halifax would reverse this difference, which would then be in favour of New York ; as against Shippegan, 228 miles, and against Halifax, 482 miles ; showing a marked advantage over Halifax in favour of Shippegan, where there is one of the finest harbours in the world, and only twenty-seven miles further off from Liverpool. But on the Pacific, the harbour at Bute Inlet is open all the year round ; whilst, according to Professor Maury, "The trade winds place Vancouver Island on the wayside of the road from China and Japan to San Francisco so completely, that a vessel trading under canvas to the latter place, would take the same route as if she were bound for Vancouver Island. So that all return cargoes would naturally come there, in order to save two or three weeks, besides risk and expense." This circumstance confirms what has been so often repeated within the last few years, of the necessity before long of more than one road across the American Continent ; at the same time that it favours the construction of a British one, which, besides the above advantages, and having its terminus in summer at an equal distance from Liverpool with New York, and 550 miles nearer in winter, would possess that of a temperate zone.

This is so true, that besides the Central Pacific, the Americans, with their accustomed energy and activity, are preparing the construction of a North Pacific railroad. This line will run from Lake Superior, along the Upper Missouri and near Fort Benton, to Seattle on Puget Sound, a distance of 1,775 miles ; and the total length across the Continent from New York, *via* Chicago and St. Paul, will be 3,124 miles, or 237 less than by the Central Pacific. Unless a counterline be built through British territory, this road will furnish the only outlet to the Red River Settlement and Saskatchewan territory, and thus prepare the way for their separation from the Mother country.

DIFFICULTY OF CONSTRUCTING A RAILROAD THROUGH AN UNSETTLED COUNTRY.

Such difficulties do not deter the Americans. With them, on the contrary (as they have learnt from experience), settlement and the institutions of civilization not only follow, but it may be said actually accompany the construction of a railroad. And such would undoubtedly be the case in British America, provided we set about it in good earnest, so that the fertility and beautiful character of the country to be settled were generally known. Let *that* be published in England, in Ireland, on the Continent, everywhere, and made known constantly, vigorously, and there will be no want of emigration. Let fair inducements for emigration be held out to our industrious poor, by liberal grants of land and other assistances, and you will have no want of respectable emigrants from this country ; to say nothing of the very considerable British emigration, which would set in from Canada and other points of the American Continent.

But we were told in Parliament, " Canada is not yet fully settled, and remains partially unoccupied, so that it is not likely that persons possessing the necessary resources to invest in land would push into regions far beyond." But this is a complete fallacy, for emigrants are doing it every day, and pushing on from Canada, as we unfortunately see, towards the back States of the Union. And the reason of this lies not so much in the old objection about the severity of the climate and the expense of clearing, as in the notorious fact, that all the good lands in Canada within reach of the present communications have been taken up ; so that those left in the market and for sale are of inferior quality, and those in private hands too dear. In the same way in the Eastern States, the days are passed when an emigrant with £100 could buy a farm which would enable him to become a prosperous man ; and he has now to push further west. The Saskatchewan territory, however, could be easily opened to emigrants, and a summer communication established from Lake Superior to Fort Garry, by the line laid down by Mr. Dawson, and already mentioned. Beyond that point, the North-West Territory has been gifted by nature with water communications of the very first order, which will not only become invaluable at a future day for colonial inter-communication, and transporting the farming produce of the settlers ; but, pending the construction of a railroad, would only require a few connecting links to make them available, so as to offer an easy mode of conveyance during seven to eight months in the year across the whole Continent, and that at a moderate cost. The writer has carefully studied the details of such an Overland communication, and put them into a tabular form, with the exact distances, altitudes, and other general information. It will be found in the Appendix (D), and whatever may be the result as regards the traffic for an Overland Route, it shows how greatly these natural channels must contribute to the future development and prosperity of the country. Much has been said (though chiefly by interested or biassed parties) about the rapids of the Upper Fraser and the shoals on the Saskatchewan, which would be such as to render their navigation impossible for steamers. The former have been much exaggerated, besides which, the two worst can be easily avoided, (as is shown in the Appendix, pages 43-4). As to the latter, it is admitted that the great Saskatchewan, from Lake Winipeg to the Forks, is perfectly navigable, a distance of 371 miles; and if, in the course of the next 600 miles up the North Branch, there exist a few shoals, there are also navigable stretches between, of 60 to 100 and 150 miles. Besides there is every reason to believe that these shoals could be easily deepened. Let the Americans get possession of that magnificent river, and, as on the Fraser in 1858, which till then had been declared to be unnavigable, the steam whistle would soon be heard along the banks of the Saskatchewan, from the confines of Lake Winipeg to the foot of the Rocky Mountains.

RIGHTS OF THE HUDSON'S BAY COMPANY.

There can be no doubt that, in order to establish an effectual communication through British North America, the Saskatchewan Territory must be thrown open to settlement; a thing which would have been done long since, had it not been for the obstacles placed in the way by the claims of the Hudson's Bay Company. These are now so far modified, that the Company have at last made up their minds no longer to struggle with the march of events; and are willing to cede the territory, provided they receive, as they say, a fair compensation. Latterly, the Home Government has requested them, and they have consented to make a written proposal to that effect to the Canadian Government. On the other hand, their title to the whole country under their rule has been contested on various grounds; but more particularly that to the Saskatchewan Territory, the waters of which fall, it is true, into Hudson's Bay, but which, it is asserted, may have belonged to France, but certainly did not belong to England at the date of their charter. Such is the view almost universally taken on the subject in Canada; whilst the Home Government, who are becoming anxious for a solution, but consider the matter as more directly concerning the Dominion (now that the latter has been established) have referred the Company to it for the cession of the Territory, or at least a part of it, and for subsequent compensation. These difficulties are again complicated by the question of sovereign rights and jurisdiction, which no private party can yield with propriety, and which are at present so ineffectually exercised by the Company at the Red River Settlement; by the unwillingness of the Canadians to acknowledge the title of the Company, still more to pay for it, and the large sum required; and, finally, by the vast extent of territory proposed to be ceded, and which is far too great to be annexed to Ontario. Besides, it would only, or nearly only, benefit that province, whilst New Brunswick and Nova Scotia would have to pay their share of the price.

How all these conflicting difficulties will be arranged, and a compromise speedily arrived at, it seems difficult to say. Since negotiations are pending, if a certain amount of money be all that is requisite to settle the matter, it seems fair that the Canadian Government, who would be most benefited by the cession, and who could dispose advantageously of the lands made over to it, should also find the means; especially if England agreed to contribute anything towards an Overland railroad. But one thing is certain, namely, that the Government must show a determination to remove this *real difficulty*, so that the question may be settled as speedily as possible, before a road of any kind can be opened across the Continent.

POSSIBILITY OF DIFFICULTIES WITH THE UNITED STATES.

This in plain English means, that supposing a railroad to be built, Canada is so much at the mercy of the chapter of accidents, that England might not even retain the use of it. And here we enter on a different phase of the question, and rather a delicate one; for without wishing to attribute to the American people any positive feeling of enmity towards this country, which certainly does not exist among the more educated classes, or in the rural districts, it is impossible to deny that in the larger towns, where the poorer and labouring classes are for the most part of recent foreign origin, and where the Irish element more especially abounds, the feeling against England is strong. The latter hate us cordially, as we all know, and as they are the greatest political intriguers in the country, their opinions are listened to and often made use of by those who do not entertain them. Add to this, that the Americans themselves are becoming more and more ambitious as their power increases, until many of them have come to believe that the whole American continent, from the North Pole to the Isthmus of Panama, ought of right to belong to them; and we shall be able to account for the aggressive tone which often displays itself in the public prints, but which is certainly much more exaggerated than real.

That Canada presents a long line of defenceless frontier may be true, but the invasion of a country is one thing, and its conquest another; and as long as the Canadians remain as loyal to England as they now are, and are likely to be, and have been for a century, there is little danger of any such event. The Americans have attempted once or twice to overrun the provinces, but they have always been beaten, and the struggles of 1812, 1813, and 1814, prove that the Canadians can defend every inch of their territory, and drive back their enemies when attacked in their homes. Besides, as regards the line of frontier, there are certain vital points in Canada, of which an enemy must acquire possession before he could obtain any decided footing. These are few in number, and if the arrangements now under consideration are carried out for their defence, and the inter-colonial railroad about to be begun completed, so as to connect the different provinces with each other, and more especially with the Atlantic in winter time, the militia and volunteer forces of the country could resist any attack on their own ground with the best possible chance of success. The weakest points would, of course, be those where there are fewest settlers, or where the country is almost unoccupied, as in the case of the Red River Settlement and the Saskatchewan Territory. So that the settlement of the latter, far from "weakening our military position," as was asserted not long since in Parliament (as if an increase of population could become a source of weakness), would remove at once both the pretext and the facility for invasion. I say the pretext, for it is well-known how much the Americans envy us the possession of that Territory; and justly enough, since we do nothing with it. That such a beautiful country should be made a preserve for wild beasts, and converted into a wilderness, when millions of our countrymen, who are without a home of their own, would be too happy to have a few acres, and *might* have them with a little assistance; whilst the settlers who occupy the smallest portion of it at the Red River Settlement are so neglected and cut off from the rest of the world, that they are obliged to burn their corn for want of a market, is truly a sin before God and man; a stain upon our national character. That this standing reproach should apply more directly to the Home Government may be true; but when a territory like that of the Saskatchewan is placed by Providence in the hands of an enlightened and powerful nation like the English, such neglect on their part becomes a breach of trust, and a dereliction of duty; and we deserve to be deprived of it. As Mr. Howe of Nova Scotia remarked latterly, when, after contrasting the progress of the Western States with the unoccupied condition of that noble tract of country, he asked, "What has England ever done with that territory?"

But to return from this short digression; Canada's best defence (again to quote Mr. Howe's words) is, "That a feeling of profound attachment to British institutions pervades every portion of Canada, including all races, religions, and parties; and that they are all animated with the conviction, that under the free institutions that prevail in British America, there is a security for life, and property, and personal freedom, which is to be found neither under the despotic governments of Europe, nor the Republican institutions of America." And these feelings have been strengthened by the late Union, which has given the country a higher *status* than could ever have been enjoyed by any province separately, and has elevated the Canadians in their own opinion and in that of the world.

Then, again, if we cast our eyes on the present condition of the United States, are they, or will they be for many years, in a position to become aggressive? or rather is not the time drawing on when separation and discord will again become inevitable? It is true they are still nominally united, if a re-union that has been achieved at the cost of the lives of half-a-million and the tears of the vanquished, can deserve that name. And time, it is true, will gradually calm the tempest of angry passions, and possibly put an end to the internal convulsions which at present paralyze the country. But for the present, a Republic, one half or the better half of which, has after a protracted struggle, succeeded in conquering and subjugating the other, and now rules over it with more than military despotism; with the West on her flank discontented, the South in her rear ruined and prostrate, but desiring separation as unitedly and more fervently than ever; and a debt of three thousand millions of dollars, bearing a heavy rate of interest (which will no doubt be paid in the course of time, but which, in case of war, could neither be repudiated nor allow of fresh loans), is not in a position, at least for the present, to become aggressive; unless from some very justifiable cause of provocation. But no such cause exists at present, nor is there any good reason why on either side any such should arise.

Canada, on the other hand, is strongly opposed to any union with the Republic. There is little sympathy in Canada with institutions, the defects in the workings of which, however ignored by superficial admirers at a distance, or slurred over here for political purposes, are far too apparent to such near observers as the Canadians. On the frontier men are in daily intercourse with their neighbours, and it is proverbial that the nearer the frontier, the greater the aversion to live under American rule. Free from the elements of discord and the other sad consequences of a prolonged civil war, from the burden of heavy taxation, from the curse of political hacks and intrigues, and the most glaring political corruption, the Canadian stands aloof; and proud of the real liberty he enjoys, compares the superior advantages of his position with that of his Republican neighbours. As long, therefore, as the conduct of England towards Canada remains liberal and conciliating, it will be warmly reciprocated, and doubly repaid by the friendly and increasing commercial intercourse of half a continent; whilst any attempt at aggression on the part of the United States would instantly convert the Canadians into the bitterest of enemies; and if persisted in, might bring on, in another form and under different circumstances, a species of civil war, with a repetition of the scenes of mutual extermination which decimated the South. May we trust that nothing so fearful will ever come to pass.

THE GREAT EXPENSE OF THE UNDERTAKING.

The cost of an overland railroad with 4ft. 8½in. gauge,* and a single line of rails, from Ottawa to the Pacific, would be, from what we have seen of the nature of the country, and with no land or lumber to pay for, no fencing or parliamentary expenses, and provided the extrava-

* A 3ft. 6in. gauge would be much cheaper. It would greatly diminish the cost of the roadway, the rolling stock would be less expensive, and the working expenses lighter. In countries like Sweden, Norway, and Australia, with a sparse population, and great distances between, it has been found to work admirably, and the writer would decidedly recommend its adoption.

gance so common in the construction of English railways be carefully avoided, relatively small. Allowing ten per cent. for side ways, and comprehending station accommodation, engineering expenses, rolling stock, reserve fund, and contingencies, it may be roughly calculated as follows:

	Miles.		Dlrs.		Dlrs.
From Ottawa to Fort Garry (nearly level)* ..	1165	at	50,000	=	58,250,000
From Fort Garry to Jasper's House, foot of Rocky Mountains (level plain)	1100	at	40,000	=	44,000,000
From Jasper's House to the Head of Bute Inlet (partly plain)	620	at	45,000	=	27,900,000
Total	2885				130,150,000

Or say twenty-seven millions sterling, exclusive of interest during the construction and until the line be in activity; of which twelve would be for the portion from Ottawa to Fort Garry, which would open an immediate communication through British North America all the year round.

We shall be told, of course, that such an outlay is far too great to be thought of. But what we have to consider is not merely the amount, but the object to be attained, and whether that is commensurate with the outlay. If the commercial supremacy of England is at stake, and that has been pretty clearly shown, what are seven and twenty millions, as compared with the sad downfall which must inevitably follow such a loss, and the decay and ruin of our country? Never was so large a sum of money more usefully, more wisely applied; and in vain might we ransack the history of our national debt to find a parallel. In times past a single subsidy to some Continental potentate has often cost more.

The nation, however, might be spared any such outlay if a company could be found to undertake the work; a thing which could most likely be accomplished by offering liberal grants of land which are at present of no value, but which in the Western States have in several instances paid the whole cost of the railroad; by engaging to subsidize mail steamers in connexion with the line as soon as required, on the Atlantic and the Pacific Oceans, instead of as at present to New York; by authorising the company to issue mortgage bonds to a certain amount; and by paying the interest as a bonus, or encouragement (so as to diminish to some extent the risk, if any, and the large amount of capital required), until the road was completed and became self-paying; which, as will be presently shown, would be the case before long. Four per cent., on a gradual outlay of twenty-seven millions spread over six years, would amount to about four millions and a half, or less than the sum about to be advanced for the acquisition of our telegraphs, or that expended on the Abyssinian Expedition. And if the latter has added so much to the grandeur and prestige of the English name, there can be no reason why a similar amount should

* The 115 miles of direct road from Montreal to Ottawa are not included in the calculation, because this improvement can be deferred for the present.

not be readily granted, when the object to be attained is pregnant with such infinitely greater consequences. But Canada, who would gain so immeasurably by the undertaking, should also contribute her share; in which case, the sacrifice would be as trifling for England, as it would be temporary for both countries. I leave these considerations to the statesman who may hold the reins of government when Parliament again assembles. Without, perhaps, being aware of it, the commercial destinies of the country will then be in his hands; and I will merely add, that he may not only immortalize himself by bringing forward and accomplishing such a measure, but that he would have the support of the whole nation, if once made to understand the issue of the case, and that the future of England depended on it.

SMALL PROBABLE TRAFFIC AND CONSEQUENT RETURNS.

It has already been observed how quickly an American railroad in the Western States is followed, or rather accompanied, by settlement and civilization. This will be better exemplified by the following figures, showing the astonishing increase in the earnings of some of the Western railroads in the course of the last four years :—

	1863.	1867.
Chicago and North Western ..	.. 2,811,544 Dolls.	.. 11.532,348 Dolls.
Chicago, Rock Island, and Pacific	.. 1,959,267 „	.. 4,153,281 „
Michigan (Southern) ..	.. 3,302,543 „	.. 4,613,754 „
Toledo, Wabash, and Western	.. 1,439,798 „	.. 3,784,816 „

The fact, however, of an intermediate, unsettled country, like that to be traversed by the proposed line, and the consequently small amount of "way" traffic to be expected in the beginning, would be more than counterbalanced by the "through" traffic, and the daily increasing crowd of passengers, who, homeward and outward bound, would cross the continent.

The following more or less authenticated facts, from the Report of the Union Pacific Railroad Company, with the corrections and modifications introduced by the writer, may give some idea of what this traffic would most likely amount to.

MERCHANDISE.	Tons.
Ships from the Atlantic round Cape Horn, 100 at 800 tons* of goods each	80,000
Steamers connecting at Panama with California and China, 55 at 2,200 tons. Say 40 with 1,500 tons of goods each	60,000
Overland Trains, Stages, Horses, &c., 30,000 tons, say	20,000
	160,000
N.B.—Before the construction of the Panama railroad, 27,000 teams left two points on the Missouri for their westward journey in one year.	
Return freight as much more, say only the half	80,000
Instead of 460,000 tons..	240,000

* The ton is the American one of 2,000 lbs.

PASSENGERS.

110 steamers both ways, at 454 each, say 80 at 625 each	50,000
(They often carry 1,000 passengers and more, seldom less than 600)	
200 vessels both ways round the Cape, at 20 passengers; say 15 ..	3,000
Overland both ways, 100,000; say	87,000
Instead of 154,000	140,000

RECAPITULATION.

American Line from Omaha to San Francisco (1830 *Miles*).

	Dolls.
460,000 tons at 1 Doll. per cubic foot, or 34 Dolls.	15,640,000
(The present price from New York to San Francisco is, by Panama, 70 Dolls., or £14 8s., and by Cape Horn about 12 Dolls., or £2 10s.)	
154,000 passengers at 100 Dolls.	15,400,000
(The steerage price by Cape Horn is about the same, or £20, whilst the passage by Panama costs from £30 to £50.)	
	31,040,000

This sum is then swelled in the American calculation to 55,200,000 dollars (£11,381,443) on the strength of the expected increase in the number of passengers, which it is supposed would be doubled, and who would be charged a higher price (150 dols.); whilst the freight is reduced to 300,000 tons. Deducting one half for running expenses, there would remain net £5,690,700 for 1830 miles.

British Line from Ottawa to Bute Inlet (2885 *Miles*).

Supposing the "through" traffic on this line to amount only to one half of the present traffic between the Eastern States and the Pacific, with the deductions made by the writer, and the prices to be, as on the American line, $1\frac{7}{8}$ cents per ton per mile for goods—£11 3s. (say £10) and $7\frac{1}{2}$ cents per mile for passengers, making £44 12s., which we will reduce to 5 cents, or £30; and we shall have:—

One-half of 240,000 tons, at £10	£1,200,000
Ditto of 140,000 passengers, at £30	2,100,000
	£3,300,000

Deduct the half for running expenses (which are always comparatively smallest on the longest lines of road) and there will remain £1,650,000 —against the American calculation of £5,690,700 for less than two-thirds of the same distance. The above figure would at once give a dividend of six per cent. on a capital of twenty-seven millions. But nothing has been reckoned for the sale of lands, which would alone form a most important item; nor for the carriage of mails (laden with the correspondence of half the globe), nor for that of the precious metals; or for the "way" traffic (during and after the construction of the road) with the Cariboo gold mines, and the Red River Settlement. This latter would soon become important, were it only by transporting the produce of the plain, in return for lumber and fuel from the forests of the Winipeg Territory; to say nothing of the indirect trade that would immediately spring up with Lake Superior, and which would be tapped by the line. When, therefore, we take these additional elements into consideration, together with the very moderate estimate of the probable "through" traffic, there can be little doubt, not only that the line would quickly become self-paying, but that (without attempting to reach the expectations of the Union Pacific Company, which suppose a return of fifteen million dollars on an outlay of eighty-five, or $17\frac{1}{2}$ per cent), the dividend on the British line would soon approach nearer to eight or ten per cent. than six.

Any calculation, however, as to the probable returns of this Overland Railroad must necessarily be of a vague character. It has even been questioned whether any goods, excepting the very lightest and costliest, can be carried across the continent at such rates as would produce any very great disturbance in the present channels of commerce. But in a large number of instances the rapidity of transit will counterbalance the higher rate of transportation. Speed is the great "desideratum" of the day, and the best proof is in the astonishing amount of freight passing over the portion of the American railroad which is already finished, though it has then to cross the Desert. The shorter route through British territory would undoubtedly command the largest share of trade between Europe and Japan; and there can be no doubt, in a general point of view, of the vast development of trade and intercourse which must accompany the opening of these great public thoroughfares. When, therefore, we think that the distance to Sydney from Vancouver Island is, as contrasted with Panama, as 7,200 to 8,200, or one thousand miles less; that the distance between Liverpool and Shanghai by this route will not exceed 10,400 miles, being less by 4,000 than by the Cape, and 3,600 miles less than by the Isthmus of Panama; that the time from London to Hong Kong would be reduced to about forty days; and that the English trade to China alone amounts to thirty-eight millions sterling; it is easy to foresee what amount of traffic would soon be running over this "great highway of nations," with seven hundred millions of consumers in Asia at the terminus—a traffic sufficient to occupy a fleet of first-class steamers on either ocean.

Nor have we made any mention of an Overland Telegraph. The correspondence by telegraph between Victoria, Vancouver Island, and the gold mines in the north, and between Victoria and San Francisco, in connection with New York and the East, is already considerable, and would of course be vastly increased by an Overland Telegraph. The telegraph which crosses the Desert, from the Missouri to San Francisco on the Pacific, paid more than the cost of its erection the first year; and though the circumstances are in some respects very different, telegraphic communications are as necessary to our commercial relations as railroads; and there can be little doubt, that the proposed line would give large and increasing returns, the instant it could be connected with Canada, and consequently with England. This might be done in less than two years; when an uninterrupted communication, under British control, would be established between England, Montreal, and British Columbia, by the telegraphic wire; and thence later across the Pacific to Japan and China.

ANTI-COLONIAL THEORIES.

These, one would think, should be banished from the discussion of such a national question, as being foreign to it. And yet I am told by certain apostles of this school, that if trying to bring about so great a national work (which they almost deny, for they seem to think, that if the Americans are outstripping us, we must let them do so rather than spend a halfpenny abroad), it is simply because the Colony of British Columbia, whose interests I happen to represent in a certain measure, will be benefited by it. Of course it will, and so will Canada, and so will England ten times more. "But it is in the nature of things," they say, "that British Columbia, and the trade and control of the Pacific, with all its consequences, should belong to the United States." More than that, "We might perhaps be taxed in order to keep them; and, therefore (though rather annoying), we had better make up our minds to give them up at once." But if such conclusions are worth listening to, England is also *in the nature of things*, and of itself, a small unimportant island. In which case, our forefathers have been working strangely against nature for the last two hundred years, and acting very foolishly in trying to add to it those foreign possessions which have made it what it is. We used formerly to be taught, that England owed her greatness and prosperity to these possessions; but this doctrine has been abolished by these gentlemen, and in the face of the most convincing facts to the contrary, the fashion with them now is, to deny that we derive our present prosperity from any such source. Many of them even go further, as is well known, and indignant at the thought of any new expense, maintain that England without colonies would be more prosperous than with them. The conclusion is, to say the least, singular; and shows how the reasoning powers of over-clever men may become perverted, and their notions gradually contracted, by continually taking the same narrow view, and only reasoning on one side of a question. Fortunately, such theories are not those of the great majority of the nation.

But whatever may be said about the cost of our colonies to the mother country, Vancouver Island never cost her one cent, unless it be an old bunting given or lent it in 1846, on the day when the island was proclaimed a Colony. On the contrary, she has been annoyed in every way by the claims of the Hudson's Bay Company, and was even held in pawn by her for some time! British Columbia yields over £600,000 of gold yearly, and would yield the double or the quadruple if her mines were more accessible; and of this sum she takes back probably one half in English goods, often of inferior quality. (Mark this, ye anti-colonists !) And yet her existence as a Colony has been as good as ignored; excepting to send out Governors, of whom the Colonists never heard, with exorbitant salaries, but without local knowledge or experience, who when there, ride over the feelings and wishes of the inhabitants, and whose sole occupation seems to be to find out fresh sources of taxation. British Columbia has not even Postal communication with the rest of the world; and however unseemly it may appear, that a great nation like England should be indebted to a foreign Power for the carriage of her Government dispatches, their arrival in British Columbia, in the absence of any Postal agreement, depends on the good will of the United States Government, and their postage is paid for by the Colony.

Now, if, as the anti-colonists will have it, the Colonies cost more than they are worth, and England would do better to get rid of us, let us be told so at once, instead of neglecting and coquetting with us by turns. If England is blind to the value of British Columbia, both in a commercial point of view and as controlling the Pacific, the United States are wide awake on the subject; and whatever the loyalty of British Columbia, if the ties that bind her to the mother country are burdensome to both, let them be torn asunder. But, if otherwise, such language as the above can only tend to indispose and alienate a Colony, which has already too many just causes of complaint. The grievances of British Columbia were submitted by the writer not long since to the House of Commons, in a petition, an abstract from which can be consulted in the Appendix (E).

POLITICAL AND IMPERIAL SIDE OF THE QUESTION.

We have now examined the principal objections which have been made to this truly national undertaking; and we have answered, and, as we think, refuted them all; with the exception of that relative to the Hudson's Bay Company, which can only be solved by the Government. But besides the many reasons already given in favour of this scheme, which has been shown to be neither visionary nor hopeless, there are imperial reasons of the greatest weight. First among which may be placed that of connecting British Columbia with the Dominion, so as to retain both of them permanently under the British flag. We have just alluded to the way in which British Columbia has been neglected, and the consequent state of disaffection there; but in a military point of view, and to quote a letter of Professor Maury's, written nine years ago, on the commanding geographical position of Vancouver Island, in connection with the different routes at that time under discussion for an Overland Railroad, he says :—

"Vancouver Island commands the shores of Washington and Oregon; and whether the terminus of the Northern (American) road be on Paget Sound or at the Mouth of the Columbia river, the munitions sent there could be used for no other part of the coast, for Vancouver overlooks them. They could not, on account of Vancouver in its military aspects, be sent from the Northern terminus to San Francisco and the South; nor could the Southern road—supposing only one, and that at the South—send supplies in war from its terminus, whether at San Diego or San Francisco, by sea either to Oregon or Washington—Vancouver would prevent, for Vancouver commands their coasts as completely as England commands those of France on the Atlantic. So complete is this military curtain, that you never heard of France on the Atlantic sending succours by sea to France on the Mediterranean, or the reverse, in a war with England. The Straits of Fuca are as close as the Straits of Gibraltar." Here is the opinion of an American, and a most competent person, who judges the case from a far higher point of view than some of our English statesmen :—

But what would become of the Dominion and of her loyal feelings towards the mother country, if after being elevated by England almost to the state of an independent nation, she were to be all at once deprived by our neglect of this communication with the Pacific, as well as of the intervening Saskatchewan Territory, 'both so essential,' as I wrote not long ago, 'to her development, to her maritime prosperity, her independence, nay, to her very existence. The interests of Canada and British Columbia, however identical with those of the mother country (a thing which England will find out one of these days), are generally overlooked or neglected in this country. And yet British America is one in interest, and together with the mother country, must be one in purpose, if the danger with which both are menaced is to be averted.' And for that purpose, the different provinces of British North America must not only be politically united, and that speedily, so as to form a whole; but must at the same time, and in a commercial point of view, be more directly and intimately connected with each other and with the mother country through regular steam communication. By these means British influences would be naturally fostered and maintained, and immigration from the home country promoted; until a friendly but independent power could be gradually developed in British America, which would not only be no longer at the mercy of the neighbouring Republic, as some pretend, but would, on the contrary, form an important counterpoise to that of the United States, and an additional guarantee for the peace of the world.

Nor is there anything far-fetched in such a prevision, which is fairly justified by the astonishing progress which Canada has made within the last twelve years (see Statistics in Appendix F); a progress greater in proportion, both morally and materially, than that in the United States. In travelling through Canada one feels at every step that she must become a great nation, in spite of all obstacles; and at the same time different in its origin, its associations, its feelings, and character, from that of the United States. Nobody can estimate the value of such a political element, or what such a country may become. As long as that counterpoise on the American Continent existed, the power of the Republic would be broken, whilst England would be mistress of a surer and more direct road to the East than that by the Isthmus of Suez, or any other she could possess. But let that weight be thrown into the opposite scale, and the rule of the United States extended over British America, and the balance of power is gone. With North America, England would lose the West Indies, and be stripped of every point on the Atlantic and Pacific Oceans; her commerce and prestige would be destroyed; her very security (with hostile armaments brought a thousand miles nearer to her coasts) endangered; and the peace of the world made a problem, dependent on the good-will or the caprice of the popular assemblies of the United States.

CONCLUSION.

It has now been convincingly shown that the best and easiest line of communication to the Pacific across the North American Continent is through British territory. In a late debate on the subject in the House of Commons, and in reply to Sir Harry Verney, who had insisted that the honour, interest, and duty of England alike required that she should take immediate action in the matter, the Under Secretary of State for the Colonies, said:—"He entertained no doubt that ultimately it would become the great thoroughfare of the world to the West;" but (alluding to the opening of the Saskatchewan Territory), "there was not yet sufficient appreciation of its value in the public mind, to cause the pressure, that he believed would yet be exerted, to be put upon the Government to bring about a settlement of the question." In other words, it was the duty of a Constitutional Ministry, though convinced themselves, to await the pressure of public opinion, before bringing forward such an important measure. But if England had a strong Government, instead of so many Heads of departments; with an enlightened statesman at the head, who not only understood the interests of the nation abroad, but had foresight and energy enough to take things in time, instead of waiting for expressions of public opinion till it is too late, we should hear a very different language.

The fault then lies with the nation at large for having no such representative. The fact is, that England, whilst slumbering under the lethargic effects of prosperity, seems not only to have forgotten that it is to our numerous colonies, our possession of the Indies, and the control of the trade between Europe and Asia, that she owes her wealth, and her existence as a great nation, but she seems to think that these must last for ever, without any further effort to retain them. In England every one is so much absorbed in his own affairs, and so habitually ignorant on colonial matters, that if he has, perchance, heard of this Pacific Railroad, he neither thinks about it, nor cares about it; still less has he reflected on its consequences; nor, in the confusion of his ideas, does he believe that the construction of a rival road can be anything more than a colonial question, and, therefore, the sooner got rid of the better. People abroad, as is often the case, take a more general, and, therefore, more correct view of the question; and the following extract from an able periodical, the "*Revue des Deux Mondes*," written nearly eight years ago (when we Englishmen were fast asleep, as we still are, on the subject), shows the importance attached to the question, even at that period, by a people little interested in it.

"England and the United States are both of them fully sensible, that the time has arrived when the sceptre of the commercial world must be grasped and held by the hand of that power, which shall be able to maintain the most certain and rapid communication between Europe and Asia. It is not merely by the Isthmus of Suez and the Red Sea that henceforth the trade with the East is going to be carried on. The Eastern continent of Asia will be waked up to a new commercial activity from other ports, and especially from the several ports of the Chinese Empire. Consequently, the empire of the world, in a commercial point of view, will henceforth belong to that one of the two Powers of England or America, which shall be the first to find means to establish a direct road across the continent of America, whereby to communicate most rapidly with the great East on the Pacific side, and with Europe on the Atlantic side. This will be the great highway by which the products of the Old World will have to be carried to the Eastern World. . . .

"Hence it is that the victory, which is to give the empire of the world, will be gained by that Power which shall be the first to establish the line of railroad across regions and countries which are yet unknown and unexplored. The struggle for the attainment of this great victory is well worth the trouble and expense which it will cost; for the empire of the seas and the commercial dominion over the whole world are the great stake which are being played for."

But the struggle which was thus predicted has now become imminent; the little cloud, then no bigger than "a man's hand," has gradually grown larger, darker, and more menacing, and the day is fast approaching, when that envied trade with the East, "the diversion of which has marked the decline of empires" is about to be wrested from England, unless she hasten to parry the blow. In the meanwhile, the high road over which this race is to come off between the two greatest commercial nations of the world, with Europe for spectator and Asia to hold the stakes, is still open to the competing parties. The vantage ground is even in favour of England, as we have shown; but while the latter has been in a state of somnolence, her active rival has been wide awake, and has laid so much of the race ground behind her and got so far ahead, that unless England strain every nerve to regain the lost distance, she will come in second best.

And can it be that in presence of such a stake England will remain inactive? that without contest, without even thinking of attempting a contest, and simply on the "laisez faire" principle, or what is worse, in order to avoid the possible outlay of a few paltry millions, England will prefer such a short-sighted, pusillanimous policy to action, and quietly consent to descend from the rank she holds among nations, and become a second-rate Power? We think not; we trust not; and if she has hitherto neglected this vital question, we may console ourselves at least with the hope, that ere long, when a new Parliament shall have infused fresh wisdom into the councils of the nation, and our statesmen, instead of wrangling over the dry bones of worn-out party politics, have been compelled to consider the matter more seriously and bring before the nation the pressing exigencies of the case, England will meet them with her accustomed energy.

I have done my best to warn her; the above scheme has occupied the whole of my time and attention for some years, and is now the object of my journey to England; and if crowned with success, I shall regret neither the time nor the expense, nor the many annoyances I have had to encounter in trying to forward it.

ALFRED WADDINGTON.

Tavistock Hotel, Covent Garden,
September 17th, 1868.

APPENDIX.

A.—REPORT TO THE NEW YORK CHAMBER OF COMMERCE.

"There is in the heart of North America a distinct subdivision, of which Lake Winipeg may be regarded as the centre. This subdivision, like the valley of the Mississippi, is distinguished for the fertility of its soil, and for the extent and gentle slope of its great plains, watered by rivers of great length, and admirably adapted for steam navigation. It has a climate not exceeding in severity that of many portions of Canada and the Eastern States. It will in all respects compare favourably with some of the most densely peopled portions of the continent of Europe. In other words, it is admirably adapted to become the seat of a numerous, hardy, and prosperous community. It has an area equal to eight or ten first-class American States. Its great river, the Saskatchewan, carries a navigable water-line to the very base of the Rocky Mountains. It is not at all improbable that the valley of this river may yet offer the best route for a railroad to the Pacific. The navigable waters of this great subdivision interlock with those of the Mississippi."

B.—EXTRACT FROM A PAPER READ AT THE ROYAL GEOGRAPHICAL SOCIETY, MARCH 9TH, 1868.

ROAD THROUGH BRITISH COLUMBIA.

The Colony of British Columbia is to a great extent occupied by two ranges of mountains, running N.N.W, but gradually diverging from each other towards the north, where they enclose a vast plain, of which more will be said hereafter. That on the east side bears the name of the Rocky Mountains, and the other that of the Cascade or Coast range. They have one feature in common, which is, that their eastern edge rises in both cases abruptly from an elevated plain; and in the Rocky Mountains the highest crest or ridge is also on that side; whereas the descent on the western slope, though greater, is extended over a wider distance, and, therefore, in general more moderate.

The Main crest of the Rocky Mountains, several of the peaks of which rise to a height of 16,000 feet, forms the eastern limit of the Colony, and runs from its S.E. corner at the Boundary line, in a N.N.W. direction, to beyond the Northern limit of the Colony, in lat. 60°. I say the main crest, because what generally bears the name of the Rocky Mountains is composed in British Columbia of three distinct ranges, divided from each other by rivers and deep depressions, and having each its own crest or ridge. Of these, the two western ones, though less elevated, are chiefly composed of metamorphic rocks, and, therefore, generally speaking, more distorted and abrupt than the rounded granitic peaks and domes of the main crest. The whole forms a triple fence as it were to the colony, or one vast sea of mountains, averaging from 150 to 160 miles wide.

The Middle range, which as before said, is somewhat lower than the main one, and which takes the names of the Purcell, Selkirk, and Malton ranges successively, is separated from the main ridge by the Kootanie River, the Upper Columbia, the Canoe River and the Upper Fraser; and presents one uninterrupted line of mountains, some of them 12,000 feet high, parallel to the main range, for 240 miles from the boundary line to the Great Bend of the Columbia, in 52° N. lat. The Columbia River here runs towards the north, and after separating the above middle or Selkirk range from the Rocky Mountains proper, cuts through it at the Big Bend, and turning south, again separates it in its downward course from the third or more westerly range. But the travellers who have discovered the different passes [such as they are in this latitude] through the Rocky Mountains, were unable to push their explorations further than this eastern or upper portion of the Columbia, excepting near the Boundary line; so that neither the middle range nor the western one, which were, perhaps, supposed, as being less elevated, to present less difficulties, had been hitherto examined. In consequence, however, of the gold discoveries at Kootanie and the Big Bend, or in connexion with them, they were carefully explored last year; but no practicable pass could be discovered through the Selkirk range, which thus presents an impenetrable barrier for a railroad in that direction.*

The Third or more westerly range is the least elevated of the three, though still ranging from 4000 to 8000 feet high. South of Fort Shepherd and the Boundary line, where it forms eleven sharp ridges running north and south, it bears the name of the Kulspelm Mountains, and further north of the Snowy Mountains or Gold range. The Bald Mountains in Cariboo, 6000 to 8000 feet high, are also a continuation of this range, which after crossing the Fraser, below Fort George, lowers towards the north, and takes the name of the Peak Mountains. The only good pass from the Columbia through this third range is to the south end of Soushwap Lake, and was discovered last year by Mr. Moberly, the Government Engineer, at Eagle Creek, in lat. 50·56. An important feature in both the middle and western ranges just described, is their gradual depression north of Cariboo to where the Upper Fraser, after separating the middle range from the Rocky Mountains, abandons its north-westerly course, and makes a circular sweep through the depression from east to west and then south to below Fort George. This depression forms a large tract of level, flat country on each, but more particularly on the south side of the Fraser; and as the country and climate are both well adapted for settlement, offers every inducement and facility (if indeed it be not the only pass), for a future railroad through these two ranges of the Rocky Mountains.

The Cascade range forms the Coast line of the Colony, which it follows, from near the mouth of the Fraser into the Russian [now American] Territory. Its average width is about 110 miles, and it may also be considered as a sea of mountains, some of which attain, if they do not exceed, a height of 10,000 feet. Its crest, starting from Mount Baker, a

* A scheme, it is true, has been broached and even patronized in the interest of New Westminster, for overcoming this difficulty, by making use of the Columbia for 150 miles, of which 90 North, from the Eagle Pass in the next Range and in lat. 50° : 56, to the Boat Encampment and the Big Bend of the Columbia, in lat. 52°, and then 60 miles south to Blaeberry River; from which point the road would pass north of Mount Forbes, 13,600 feet, and Mount Murchison, 15,789 feet, and by Howse Pass, 6347 feet high, over the main crest of the Rocky Mountains. But forty miles above Eagle Pass the navigation of the Columbia is interrupted by the Dalles de Mort or Death Rapids, and the formidable bluffs on either side of the river would render the construction of a wagon or railroad most difficult, supposing there were no greater difficulties beyond. Such a road may do very well on paper or to show in England, but, practically speaking, could never be carried out.

few miles south of the Boundary line, passes a little north of the head of Jervis Inlet, some 25 miles north of the Head of Bute Inlet, 22 miles east of the head of North Bentinck Arm, and crosses Gardener's channel about 20 miles west of its head. From Mount Baker the Cascade range throws out a spur east and north, in the direction of the Great Okanagan Lake and Fort Kamloops, so as nearly to join the Gold range; and it entirely envelops the Fraser, from a little above Harrison River [55 miles above New Westminster] up to its junction with the Thompson at Lytton, and even a few miles beyond, on both rivers. But the most rugged portion in this direction lies between Yale and Lytton, where mountain succeeds mountain, and where those along the river present the most formidable aspect; bluff after bluff of solid perpendicular granite, intermingled with steep slides of rolling rock, washed by a deep impetuous stream, and 1500 to 2000 feet high. In short, not only has this portion of the Fraser valley been declared to be utterly impracticable for a railroad by Major Pope and other competent authorities, but it is so fenced in with mountains, that there could be no reasonable way of getting at it with a railroad, if it were. It is over these mountains that the present wagon-road passes, at an elevation, in one place for nearly 40 miles, of 3600 feet above the sea—the only road to the Cariboo mines and the north of the Colony, and, considering circumstances, a lasting monument of Sir James Douglas' energetic and provident administration. Unfortunately, the difficulties [as may be seen in "Milton & Cheadle's North-west Passage, p. 356," where there is a good sketch of one of them] were alpine. Many places are most dangerous, the endless ascents and descents fatiguing and laborious in the extreme, and as the sharp turnings, besides many other portions, have had to be built up to a great height on cribs or cross timbers which must soon rot, the repairs will form a heavy charge on the Colony.

So that, supposing the difficulties through the Rocky Mountains to be got over, the Cascade range still intercepts all direct communication by railroad between the Eastern part of the Colony and New Westminster. To say nothing of the utter worthlessness of the greater part of the country to be traversed, amounting to over 450 miles out of the 600 from its Eastern limit by Howse Pass. Add to this, that the navigation across the Gulf of Georgia and at the entrance to the Fraser, *by a narrow, intricate channel, through shifting sands, full five miles long*, is both difficult and dangerous, and that the river itself is frequently frozen up in winter for long periods of 2, and even occasionally 3 and 3½ months; and it will be evident to every impartial mind, that New Westminster, with its 700 or 800 inhabitants, can never become the terminus of an Overland Railway to connect with Victoria and the ocean.*

Further north along the Coast, there are numerous inlets which penetrate into the Cascade range, but the greater part either terminate abruptly, like the fiords in Norway, or are too distant; or like Gardener's Channel, Dean's Canal, and the Skeena, are too far to the north-west to be available for any present communication with the mines or the interior. There are, however, two exceptions: The North Bentinck Arm, by Milbank Sound, in lat. 52°.13, and Bute Inlet, opposite Vancouver Island, with a safe

* It has been proposed latterly to substitute Burrard Inlet for the Port of New Westminster. The tide runs through the neck or entrance of this Inlet at the rate of 8 knots an hour, thus requiring a steam tug. Outside, there is a good roadstead in English Bay, though rather exposed and less secure than the harbour at Bute Inlet. A railroad could be easily built from New Westminster to English Bay, but the Cascade range intercepts any road to New Westminster, as we have just seen; so that the difficulty remains much the same.

and easy inland communication by steam to Victoria, distant 185 nautical miles. Both these inlets terminate in a valley of some extent; and as attempts have been made to open both of them, it becomes necessary to explain why the writer gave a decided preference to Bute Inlet, for a wagon road and *a fortiori* for a railroad, over Bentinck Arm or any other line.

Superiority of the Bute Inlet Route.

The advantages of the Bute Inlet Route consist: In its central position; fine townsite and harbour; or rather two harbours, accessible at all seasons of the year; its easy and safe connection with Victoria and the ocean, and the proximity of the coal mines at Nanaimo.

The harbour at Bella Coola, on the Bentinck Arm träil [the only other feasible route to the mines], is situated 435 miles futher to the north, and has been pronounced totally unworthy ; presenting no shelter, no good anchorage, no good landing place; but a vast mud flat, with a mile of swamp, intersected by a shallow river barely navigable for canoes. Or to quote the words of Lieut, Palmer, of the Royal Engineers, in his official report on the Bentinck Arm Trail: "A large flat shoal, extending across the Head of the Arm, composed of black fetid mud, supporting a rank vegetation; bare at low spring tides for about 700 yards from high water mark, and covered at high tide with from 1 to 8 feet of water, and at a distance of 800 yards from shore, terminating abruptly in a steep shelving bank, on which soundings rapidly increase to 40, and soon 70 fathoms." The whole is, moreover, subject to violent winds and powerful tides.

On the Bute Inlet Route the snow, owing to the more moderate elevation, and its more southern latitude and aspect, melts fully three weeks sooner than on the Bentinck Arm Trail; and the road is dry, entirely exempt from snow-slides, and level the whole way through. Unlike the endless mountains on the Fraser route, or the steep, unavoidable ascent from the sea, and numerous swamps by that of Bentinck Arm. The Bute Inlet Trail *cuts through* the Cascade Mountains by a deep valley studded with rich bottoms, affording plentiful pasture, and rising imperceptibly for 80 miles, when it nearly attains its greatest height (2,500 feet); from which point forward in the plain, it was free from snow for 25 miles in February, 1862. The Bentinck Arm Trail, on the contrary, is obliged to *climb over* the range, owing to the valley, when 35 miles from the Inlet, turning abrubtly to the S.S.E. and running longitudinally with the range, instead of cutting through it; so that the trail attains, in the course of a very few miles from that point, a height of 3,840 feet, as will be better shown by the following table compiled from Lieutenant Palmer's report:—

	Gradients.				
	Miles.	Per Mile.	One in	Rise.	Altitude
From the Inlet to Shtooiht, at the turn of the valley	35			say	Feet. 500
Thence to Cokelin, "by a narrow gorge, hemmed in by steep and continuous cliffs."	14	Feet 43·6	121·1	Feet 610	1110
From Cokelin to the Great Slide	5	356·0	14·8	1780	2890
From the Great Slide to the Precipice ..	11	86·3	61·2	950	3840
Or supposing it possible to equalize these grades (a thing next to impracticable) we should have	30	111·3	47·4	3340	

"After which the trail continues to rise gradually, the soil becoming shallow and meagre, the vegetation thinner and inferior, for 60 miles more, till it crosses the summit ridge at an altitude of 4,360 feet" (Lieut. Palmer's report.) And it only enters on good soil some 20 miles before crossing the Bute Inlet Trail at Benchee Lake; whereas along the latter line the bunch grass peculiar to the country flourishes over thousands of acres.

Finally, the distance from Bute Inlet to the mouth of Quesnelle river is fully 25 miles less than that by the Bentinck Arm Trail, and not much more than half of that from New Westminster (222 against 393); besides having no portages or mountains. Thus presenting an open communication during the whole winter, which exists on neither of the other routes; and a diminution of nearly one-half in the time and cost of conveyance, as compared with that by the Fraser. Lieut. Palmer in his report admits "the geographical advantages of the Bute Inlet Route over the others."

Another item in favour of the Bute Inlet Route is its great *Strategical Security* in case of any difficulties with our American neighbours. The Fraser river, from Fort Hope downwards, runs for 80 miles parallel to the boundary line, and at a distance varying from 6 to 12 miles from that frontier; whilst the only road from New Westminster to Hope and the interior has been constructed between them. So that a detachment of a few hundred men could at almost any point intercept all communication, and literally starve out the whole colony. The Bute Inlet Route, on the contrary, would be perfectly safe and its approaches impregnable.

General Features of the Ground over which the Railroad would pass from Bute Inlet to the mouth of Quesnelle River.

The valley of the Homathco river, which falls into Bute Inlet, presents a deep cut or fissure through the Cascade mountains, varying from three miles to less than a quarter of a mile in width; is 84 miles in length, and rises imperceptibly to a height of 2,400 feet or more above the sea, at the point where it enters on the plain beyond the mountains. For the first 31 miles, up to the canyon or defile, the bed of the valley is composed of diluvial soil, consisting of a sandy clay or loam, and forming a hard dry bottom. The canyon itself is exactly one mile and a quarter in length. Beyond the canyon the valley again forms and opens for about six miles, the soil partaking of the nature of the rocks from which it is derived and becoming more gravelly and of a reddish cast. The river after this is again confined to a narrow bed, but the country is more open, and the road passes for six other miles near the river along the foot of the mountains, until the valley once more opens and recovers its flat, level aspect, which it maintains up to the plain.

The rise in the valley, though apparently uniform, presents considerable variations. Thus the canyon presents a rise in 30½ miles of only 860 feet above the sea. The river then becomes much more rapid, and gives for the next thirteen miles an ascent probably of 780 feet, after which for 40 miles and up to Fifth Lake, the rise diminishes to 630 feet; beyond which there is a sharp ascent for a couple of miles more, of say 150 feet, when the summit, or watershed, is attained.

We shall thus have the following gradients:—

	Feet.			Feet.		
Rise	865	in 30½	miles—	28·36	per mile,	or 1 in 186.2
„	780	in 13	„	60·00	„	or 1 in 88
„	630	in 40	„	15·75	„	or 1 in 335·2
„	150	in 2	„	75·00	„	or 1 in 70·4
Total	2425					

The above figures must of course be considered as only approximate.

The plain consists of a deep sedimentary soil, watered by numerous lakes and small streams, and varied by occasional elevations formed of sandstone, belonging probably to the lower series of the chalk formation, and apparently owing their upheaval to plutonic action, which has hardened or calcined the rock. They form here and there conical elevations varying from 500 to 800 feet in height. Such, for instance, are Mount Palmer to the north of Benchee Lake, and several others that figure on my map. These elevations, and the low spurs or ranges of hills that accompany them, necessitate but few deviations from the straight line, and the plain in general offers every facility for the establishment of a railroad. Towards the mouth of the Quesnelle there is a gradual descent for some miles, but unattended by any difficulty; and at the terminus on the bank of the Fraser there exists a rich plateau of cultivable soil.

Agricultural Resources on the Line.

The valley above described is in general heavily timbered, but studded, as aforesaid, with rich bottoms, capable of producing any kind of crops, and offering open spots for small farms. The plain itself (the only one in British Columbia of any extent) has been admired by all who have seen it, on account of its vast pasturages and park-like scenery. Its width, where it is crossed by the Bute Inlet trail, is about 120 miles. It begins in the neighbourhood of Lake Kamloops, is bounded further North by the S.W. end of the Great Quesnelle Lake, and crosses the Fraser, increasing all the while in width, as it stretches in a N.N.W. direction, for more than 300 miles, from the Fraser to the Skeena; beyond which river it has not been much explored. It contains millions of acres of good ground, and some of the best along the proposed route, where large tracts of land are sure to be taken up as soon as the first communications are established. Some objections have been raised as to its elevation, which averages 2,500 feet above the sea in the southern part, though gradually lowering towards the Skeena, where the climate, in consequence, becomes considerably milder. But this makes it none the less valuable for grazing purposes, which will be by far the most profitable branch of farming in the country, when there are means of conveyance. At present, the cattle consumed in Cariboo are driven overland some 500 or 600 miles from Washington Territory.

Cereals can also be cultivated with success, as is fully proved by the following list, showing some of the crops which were raised last season on the Fraser route, together with the corresponding latitudes and altitudes:

	Lat. N.	Altitude	
		Feet.	
Deep Creek . . .	52:17	2255	100 acres of oats
William's Lake . .	52:12	2135	200 acres of oats, barley and wheat
Cut off Valley . .	51:10	2973	200 acres of oats, barley, potatoes, and a little wheat
Mr. Cornwall. . .	51:00	1508	70 acres of oats, barley, and 300 bushels wheat

At Benchee Lake, on the Chilcoaten plain, in the same latitude as William's Lake, and rather more elevated, but 2° more to the west, and therefore very probably identical in climate, I saw in the autumn of 1863 a small crop of oats, barley and turnips, which Mr. Manning had raised on trial, and which had perfectly succeeded; whilst some potatoes, which had been planted in an exposed situation to the south, had been frost-bitten. The Indian horses pass the winter out of doors without fodder or stabling; the best proof that the winters are not very severe.

DIFFERENT PASSES.

It remains to say a few words on the different passes which have been explored through the Rocky Mountains on British Territory; leaving out the Athabasca Pass by Peace River, in Lat. 56°: 28, as being too far north for present purposes:

NAMES OF THE PASSES.	Ridge or Divide.		
	Lat.	Long.	Alt.
	Deg.	Deg.	Feet.
1 Yellow Head Pass, from the Athabasca to the Upper Fraser	52:54	118:33	3760
2 Howse Pass, from Deer River by Blaeberry River to the Upper Columbia	51:57	117:07	6347
3 Kicking Horse Pass, by Bow River and Kicking Horse River, to the Upper Columbia, Sullivan	51:16	116:32	5420
4 Vermillion Pass, from the South Saskatchewan by Fort Bow [4,100 feet] to the Kootanie, Hector	51:06	116:15	4947
5 Kananaski Pass, from Fort Bow by Pamsay River to the Kootanie [with a short Tunnel 4,600] Palliser ..	50:45	1:1531	5985
6 Crow's Nest Pass, by Crow River to the Kootanie ..	49:38	1:1448	——
7 British Kootanie Pass, by Railway River to the Kootanie, Blakiston	49:27	114:57	5960
8 Red Stone Creek or Boundary Pass, from Waterton River to the Kootanie, [partly on American ground] Blakiston	49:06	114:14	6030

With the exception of the Yellow Head Pass in the above table, which is comparatively straight and short, and the three last which are tolerably so, but too near the Boundary line to be available, the four others describe the most circuitous routes, among a labyrinth of glaciers, and mountains covered with perpetual snow. Besides which, the approach to them over the plain by the South Saskatchewan, is for nearly one hundred miles through an arid, sandy, treeless district, forming the northern limit of the great American Desert; instead of the rich Fertile Belt drained by the North Branch, which is also the more considerable one of the two. And it is in the very latitude of this Belt, that the great barrier of the Rocky Mountains is cleft asunder, so that the road runs along this fertile zone in a direct line up to the lowest and easiest Pass, as to a natural gateway leading to the Pacific. But we have already seen, that all the southern Passes [and Captain Palliser wished it to be distinctly understood that he considered these as far from being the best that could be discovered] are intercepted further west by the Selkirk range, which presents an impenetrable barrier, and renders them so far next to useless. When, therefore, we consider their relative altitude, their necessarily precipitous nature, and the great depths of snow [27 feet or more], under which they lie buried during eight months of the year, there can be no hesitation [and such indeed is now the general opinion] in regarding the Yellow Head Pass through the Rocky Mountains, with its easy gradients and low elevation, as the only feasible one for a railroad. But the same has been shown with respect to the Upper Fraser and the Bute Inlet valley, through the Cascade range. It is, therefore, clearly demonstrated, that these passes, which connect naturally with each other, offer the best and indeed the only really practicable line for a railway to the Pacific through British Columbia.

ALFRED WADDINGTON.

C.—"The following extracts, taken from most reliable sources, will show the character of the tract of country commonly called

"THE AMERICAN DESERT."

"The progress of settlement, a few miles west of the Upper Missouri River and west of the Mississippi, beyond the 98th degree of longitude, is rendered impossible by the condition of climate and soil which prevail there. . . . The Rocky Mountain region, and the sterile belt east of it, occupies an area about equal to one-third of the whole surface of the United States, which, with our present knowledge of the laws of nature, and their application to economical purposes, must ever remain of little value to the husbandman." (Dr. Henry, Smithsonian Institution.)

"The arid districts of the Upper Missouri are barren tracts, wholly uncultivable from various causes. . . . Along the 32nd parallel the breadth of this desert is least, and the detached areas of fertile soil greatest in quantity; but the aggregate number of square miles of cultivable land amounts only to 2,300 in a distance of 1,210 miles." (Professor H. Y. Hind).

"The arid plains between the Platte and Canadian Rivers are in great part sand deserts. The sage plains, or dry districts, with little vegetable growth except varieties of artemesia, begin in the western border of the plains of the eastern Rocky Mountain slope, and cover much the larger portion of the whole country westward." (Army Meteorological Register, U. S., page 684).

"The sterile region on the eastern slope of the Rocky Mountains begins about 500 or 600 miles west of the Mississippi, and its breadth varies from 200 to 400 miles; and it is then succeeded by the Rocky Mountain range, which rising from an altitude of 5,200 in lat., 32° reaches 10,000 feet in lat., 38° and declines to 7,490 feet in lat,; 42° 24, and about 6,000 in lat., 47°. Along this range isolated peaks and ridges rise into the limits of perpetual snow, in some instances attaining an elevation of 17,000 feet. The breadth of the Rocky Mountain range varies from 500 to 600 miles. The soil of the greater part of the sterile region is necessarily so from its composition, and were it well constituted for fertility, from the absence of rain at certain seasons. The general character of extreme sterility likewise belongs to the country embraced in the mountain region." (Explorations and Surveys for a Railway Route from the Mississippi River to the Pacific Ocean, page 6.)

As regards the project of a Southern railroad to the Pacific, we find the following description of the Colorado Desert, given by the State Geologist of California:

"Its area is some 9,000 square miles, and excepting the Colorado, which cuts across its lower end, is without river or lake. It stretches off to the horizon on all sides without one glimpse of vegetation or life. Its surface is ashy and parched; its frame of mountains rise in rugged pinacles of black rock, bare even of soil. Words are unequal to the task of describing its apparent expanses, the purity of its air, the silence of its night, the brilliancy of the stars that overhang it, the tints of the mountains at daybreak, the looming up of those beyond the horizon, the glare of the mid-day sun, the violence of its local storms of dust and sand. Parts are entirely destitute even of sand, being smooth, compact, sun baked clay; other parts are covered with heaps of sand, disposed like snow-drifts in waves of 50 or 80 feet in height." (Professor W. P. Blake.)

D.—OVERLAND COMMUNICATION BETWEEN MONTREAL AND BUTE INLET, BRITISH COLUMBIA.

		Rail-road.	Steam Navi-gation.	Rise.
Sections.	Miles.	Miles.	Miles.	Feet.
1 Railroad from Montreal to Toronto	..	345	..	..
Railroad from Toronto to Collingwood, on Nollanoassaga Bay	..	97	..	..
Existing Railroad		442	..	..
2 From Collingwood across Georgian Bay to Cabot's Head	75	..	..	..
Past Cove Island Lighthouse and across the entrance to Lake Huron; along the great Manitoulin Island to the Group of Duck Islands	85	..	..	..
Thence through the Mississaga Channel, between Cockburn Island and the Head of the Great Manitoulin, along Drummond and Joseph's Islands to St. Marie River, and through the American Canal [1⅛ mile long]..	94	..	..	..
Thence across Lake Superior, to between Isle Royale at its north-western extremity and Thunder Cape [1350 feet high] into Thunder Bay and to Current River, with a good harbour, 6 miles N.E. of Fort William ..	280	..	..	..
N.B.—Lake Superior is 600 feet above the sea, according to Sir W. Logan and Keefer. The ice on Lakes Huron and Superior breaks up a little before the end of April	534	..	534	..
		Stage road.		
3 From Thunder Bay, near Fort William [situated in a fertile valley on the north bank of the Kaministaguia and one mile from its mouth] to Dog Lake, by a surveyed line ..	..	28	..	718½
N.B.—The Kakabeka Falls, on the river, enter for 182 feet, and the Dog Portage 3 miles below the lake for 347 feet in this rise.				
4 Across Dog Lake with its gently rising banks.	10	..	..	level
Up Dog River, a sluggy circuitous stream, about 80 feet wide, with flat, swampy slopes, in a valley about 1 mile wide, to the Prairie Portage	25	..	..	18
This last portion navigable for steamers by making a dam 16 feet high across the outlet of Dog Lake, at an estimated expense of £2000 (this is now in execution)	35	..	35	..
5 Prairie, or Superior Portage, over the Summit or Divide, between Lakes Superior and Winipeg, 893½ feet above the former and 1493½ feet above the sea	2¾	..	..	157
	2¾	28	569	893½

Section.	Miles.	Stage road. Miles.	Steam Navigation. Miles.	Fall. Feet.
Over	2¾	28	569	
Middle Portage, between the Dog and Savanne Rivers	¾	..	..	16
Savanne Portage, very swampy but easily drained	1½	..	..	32
Total through an easy country	5	5	..	..
6 Down the Savanne River, a meandering stream from 40 to 70 yards wide, with muddy banks and much embarrassed by driftwood, to the Lac des Mille Lacs	19	..	..	7
Through the Lake with its numerous islands and bold rocky scenery, many of them, however, containing tracts of good soil	36	..	..	4
Down the river Seine [increasing gradually from 100 feet wide and winding through a flat wooded valley] to the Little Falls at the Junction of Fire Steel River	10	..	..	37½
This last portion navigable for light steamers by a dam about 36 feet high, above the Falls.				
	65	..	65	..
7 From the Little Falls [24½ feet high] down the valley of the Seine, now bounded by low hills, of the primitive formation, to the upper entrance of Rainy Lake	..	66½	..	367
N.B.—A broken navigation for bateaux, with 5 portages, could be easily established on this portion of the river.				
8 Down the upper and narrow portion of Rainy Lake, 20 miles; then through the main lake with its rocky shores, and 2 miles beyond, down Rainy River [with 6 feet fall] to Fort Frances, at Rainy Falls, in all	50	..	..	10
The islands in this lake [over 500 in number] are mainly composed of pale red granite and chloritic and greenstone slate, and though picturesque, present a barren and desolate appearance. The lake freezes over about the 1st of December. There is a population of 15,000 Indians here.				
Portage at Rainy Falls, 171 yards, requiring two locks	..	..	..	23
From Fort Frances down Rainy River [from 250 yards to a quarter of a mile wide, and very winding], through a beautifully fertile alluvial country, studded with maple, birch, poplar, and oak, and containing at least 260,000 acres of the very best soil, to the Lake of the Woods. There are two insignificant rapids 31 miles below the Fort, which a steamer of moderate power could stem with ease	74	..	..	26½
Across the Lake of the Woods, 55 miles, and				
	124	99½	634	523

Section.			Stage Road.	Steam Navigation.	Fall.
			Miles.	Miles.	Feet.
	Over	124	99½	634	523
	thence through a navigable channel, 66 feet wide, with two small bars of loose friable slate, in all 140 feet long, to the north western extremity of Lac Plat or Shoal Lake, in all	84	..	..	7
	N.B.—The Indians grow large quantities of maize on the islands, and wild rice grows in the greatest abundance in the whole district, forming the chief sustenance of the Indians.	208	..	208	..
9	From Lac Plat to Fort Garry, near the confluence of the Red River and the Assiniboine, and 647 feet above the sea. This line has been surveyed, and a very good route can be obtained over a level and favourable country, of which the first 60 miles wooded	60	..	..	280
	And the remainder level prairie..	31½	..	..	36½
		91½	91½	..	..
	These 90 miles of road would replace 580 miles of cartage to St. Paul, where the inhabitants of Red River now get their supplies. The expense has been roughly estimated by Mr. Dawson at £22,500; and the total cost of opening the communication by land and water, as above described, from Lake Superior to Fort Garry, would probably amount to about £80,000. The Red River Settlement contains a population of from 12,000 to 14,000, and begins 10 miles south of Winipeg Lake, extending 60 miles up the Red River and 60 to 70 miles west up the Assiniboine. The land has been truly named, "a Paradise of fertility." Many farms have been cultivated for 40 years without any appreciable falling off; and as to climate, maize never fails to ripen, and melons grow with the utmost luxuriance in the open air, and ripen in August. The Red River, which is 600 or 700 miles long, is 200 to 350 feet wide, below Fort Garry, and navigable for steamers of light draft. It generally freezes up about the middle of November, or a little later, and re-opens towards the middle of April.				
10	From Fort Garry, through the Settlement, down Red River, and then through 6¼ miles of marsh, at the mouth of the river to Lake Winipeg, 628 feet above the sea	42	..	..	10
	From the south end of Winipeg Lake to its north-western extremity and the Grand Rapid	..	..	..	..
		42	191	842	865½

Section.		Stage Road.	Steam Navigation.	Rise.
	Miles.	Miles.	Miles.	Feet.
Over	42	191	842	..
2 miles beyond, on the Great Saskatchewan	255	..	..	level
	297	..	297	..
				Rise above Lake Winipeg
11 Portage at the Grand Rapid, 3 miles, with 62 feet fall, and a small rapid above [in all 5 miles], along a steep barren ridge of magnesian [upper Silurian or perhaps Permian] limestone, on the north side of the Great Saskatchewan	..	5	..	70
N.B.—This Portage, and more especially another rapid further up, above Cross Lake, might be avoided by passing from Lake Winipeg up the little Saskatchewan, the Manitonba and Winipegous Lakes, and across the Mossy Portage, which separates the latter from Lac Bourbon, and which is 4¼ miles wide; but the navigation would be most circuitous, and the distance lengthened 83 miles to little purpose.				
12 From the Grand Rapid up the Great Saskatchewan and through Lac Travers or Cross Lake to the Rapid immediately above, which would perhaps require a lock or a dam	13	..	..	18
Thence 3 miles up the Great Saskatchewan and then through Lac Bourbon, in all	53	..	..	10
Thence up the Great Saskatchewan to near Cumberland Lake and House..	115	..	..	208
Thence to the Forks of the Saskatchewan, where large beds of tertiary coal [lignite] crop out..	190	..	..	405
From the Forks up the North Saskatchewan to Carlton House [south bank]. The river here is a quarter of a mile wide, and at the lowest waters 12 feet deep. The ice sets in about the 20th of October and breaks up about the 10th of April	73	..	..	162
From Carlton House, passing the limit of the true forests at the end of about 30 miles, and then entering on the Fertile Belt, through a rich and beautiful open country to the mouth of Battle River	98	..	..	220
Thence to Fort Pitt [in the upper and middle cretaceous formation]	115	..	..	343
Thence to Fort Edmonton, on the north bank of the North Saskatchewan [300 miles below its numerous sources in the Rocky Mountains]	215	..	..	664
The north branch of this noble river, which gathers its waters from a country greater in extent than that drained by the St. Lawrence and all its tributaries, is here 250 yards wide at low water, and so far perfectly navig-				
	872	196	1139	2100

Section.		Miles.	Stage Road. Miles.	Steam Navigation. Miles.	Rise. Feet.
	Over .. .. .. able for steamboats; for which I have Sir James Douglas' authority. Indeed the Hudson Bay Company thought seriously of placing a steamer on this part of the line during the excitement of 1858-9. Above Edmonton it is navigated by the bateaux of the Company, drawing 4 feet of water, up to Rocky Mountain House, 140 miles higher; and there can be no doubt that the lower half of this distance up to the rapids, below Brazean river, is navigable for light steamers. Cumberland House and Fort Edmonton are two of the most northern points on the whole of this Overland Route. The latter is in lat. 53° 30. 2,100 feet above Lake Winipeg, and 2,728 feet above the sea. A bed of coal, 10 feet thick, of the tertiary (?) coal formation crops out here, and beds are again found cropping out on Battle River, the Pembina, the Athabasca and elsewhere, dipping towards the east. The finest wheat is raised at Edmonton, and at St. Albans and St. Ann, two settlements in the neighbourhood.	872	196	1139	2100
	From Edmonton up the North Saskatchewan, as far as its bend towards the south, a little below the Rapids and about 6 miles below the Junction of Brazean River	80	..	..	250
		952	..	952	..
13	Thence across the plain, nearly due west, and over the Pembina and McLeod Rivers, two clear shallow streams flowing over pebbly beds, about 80 feet below the plain, to the swift turbid Athabasca, a little above the Roche à Miette and Jasper's House opposite [3,372 feet above the sea]. A coach and six could be driven over a great part of this plain	..	140	..	394
14	Thence south up the Athabasca to Henry's House, at the Head of Navigation and the foot of the "TETE JAUNE PASS"	..	..	29	83
15	Thence in a W.N.W. direction up the narrow, rocky valley of the Miette, a deep, tortuous, rapid stream, 30 yards wide, and along a small tributary called Pipe Stone River, to the Summit or Watershed of the Tête Jaune Pass, 3,760 feet above the sea. This Pass is described in "Milton and Cheadle's North-west Passage by Land," 6th edition, p. 250, as follows:—"In the course of our morning's journey we were surprised by coming to a stream flowing from the westward. We had unconsciously passed the				
		..	336	2120	2832

Section.	Miles.	Stage Road. Miles.	Steam Navigation. Miles.	Rise. Feet.
Over	..	336	2120	2832
height of land and gained the watershed of the Pacific. The ascent had been so gradual and imperceptible, that, until we had the evidence of the water flow, we had no suspicion that we were even near the dividing ridge."	22	..	..	300
Total rise above Lake Winipeg..	..	..	..	3132
				Fall.
Thence across the summit, 3 miles, and along the north side of Cowdung Lake [about 7 miles long and 1 mile wide]	10	..	..	50
Then across Moose River, joining and following the Fraser for 8 miles	8	..	..	300
Then along the north shore of Moose Lake [15 miles long] through an open country ..	15	..	..	10
Then along the Fraser, partly between cliffs of slate rock, to the North Fork, and 10 miles beyond, in all 25 miles, to opposite the Tête Jaune Cache	25	..	..	900
Thence along the Fraser north to the Rapide des Fourneaux, reputed Head of Navigation	10	..	..	90
N.B.—Rich gold prospects are said to have been found about 35 miles below this Rapid.				
Total length of Tete Jaune Pass	90	90	..	..
16 From the Rapide des Fourneaux down the Fraser and past the Long Rapid to Fort George. The Long Rapid may be about 70 miles below the Rapide des Fourneaux. Some of the boulders, it is said, might require blasting when the waters are at the lowest, in order to clear the channel.. ..	187	..	..	635
N.B.—The portion of the Fraser, between Bear River and Fort George, waters a rich, open country, fully 80 miles in length, and extending many miles back on each side of the river: with a climate milder than that of Canada, and capable of raising wheat or any other kind of crops. The river itself is not less than 6 feet deep in the shallowest parts, and 500 feet wide where narrowest, and the current is slow, more like a lake than a river.				
From Fort George, past the Isle des Pierres or Stone Rapid, and the Grand Rapid, to the Mouth of Quesnelle River, 1,490 feet above the sea	93	..	..	285
The Isle des Pierres Rapid is about 20 miles below Fort George, and only awkward when the waters are very high. The Grand Rapid is 19 miles above the mouth of the Quesnelle, and much more rapid, but straight, and it is believed, on good authority, can be sur-				
	280	426	2120	2270

		Stage Road.	Steam Navigation.	Fall.
Section.	Miles.	Miles.	Miles.	Feet.
Over	280	426	2120	2270
mounted by a steamer of tolerable power. If otherwise, the road would have to be extended 19 miles up to this point	280	..	280	2270
17 From Quesnellemouth, a small rising town, S.W. across the fine Chilcoaten plain, by Chisicut, Benchee and Tatla Lakes, to the watershed and gap at the entrance of the Cascade Mountains, on the Bute Inlet route [2347 feet above the sea]	$137\frac{1}{2}$	..	..	Rise. 857
Thence through the Cascade Range, by a level valley to Waddington Harbour, at the Head of Bute Inlet	$84\frac{1}{2}$	..	..	Fall. 2347
	222	222	..	..
	..	648	2400	..

N.B.—The foregoing figures represent the distances, with all the tortuosities of the route.

SYNOPTICAL TABLE OF THE PRECEDING DISTANCES BETWEEN MONTREAL AND BUTE INLET, BRITISH COLUMBIA.

	Stage.	Steam Navig'n	Rail-road.
	Miles.	Miles.	Miles.
From Montreal to Collingwood, by Railroad	..	..	442
From Collingwood to Current River, 6 miles, N.E. of Fort William, Lake Superior	..	534	..
From Lake Superior to Dog Lake	28	..	..
Up Dog Lake and River	..	35	..
Portage to Savanne River, easy ground	5	..	..
Down the Savanne River, the Lac des Mille Lacs and the River Seine, to the Little Falls	..	65	..
Thence along the Seine to Rainy Lake	$66\frac{1}{2}$	..	..
Through the Lake, down Rainy River, and across the Lake of the Woods, to the North-west end of Lac Plat or Shoal Lake	..	208	..
Thence over the Plain to Fort Garry, Red River Settlement [with 12,000 to 14,000 inhabitants]	$91\frac{1}{2}$	..	..
Down Red River, to the North-west end of Winipeg Lake, and the Grand Rapid, 2 miles beyond, on the Great Saskatchewan	..	297	..
Portage along the North Bank	5	..	..
Thence up the Great Saskatchewan and its North Branch to below the Junction of Brazean River, 80 miles above Fort Edmonton, and the neighbouring Settlements of St. Alban and St. Ann	..	952	..
Thence to Jasper's House, Lat. 53°.12, at the foot of the Rocky Mountains	140	..	..
	336	2091	442

	Stage.	Steam Navig'n	Rail-road.
	Miles.	Miles.	Miles.
Over	336	2091	442
Thence south up the Athabasca to the foot of the Yellow Head Pass	..	29	..
Through the Pass to the Upper Fraser	90	..	..
Down the Fraser to Quesnellemouth [road to Cariboo] at the Junction of Quesnelle River	..	280	..
Across the Chilcoaten Plain, 137½ miles, and through the Cascade Range, 84½ miles, by a level valley to Waddington Harbour, Head of Bute Inlet	222	..	..
Total 3490 miles, requiring from 20 to 23 days' travel	648	2400	442

RECAPITULATION.

Existing Railroads	..	442	..
Bute Inlet Road	222	..	..
Other portions of stage road	426	648	1090
Steam Navigation	..	..	2400
Total Distance from Montreal to Bute Inlet			3490

POSSIBLE FUTURE SHORT CUT FOR A RAILROAD IN PLACE OF THE NAVIGATION BY THE UPPER FRASER.

		Rail-road.	Steam Navi-gation.	Rise.
Section.	Miles.	Miles.	Miles.	Feet.
1 From opposite the Tête Jaune Cache, South across the Fraser, then up the valley of the Cache, over easy undulating sandy ground, and across Cranberry River to the Watershed of Canoe River	14	..	[?]	240
Thence down to the bed of Canoe River, worn to a considerable depth in the sandy soil ..	2	..	[?]	Fall 150
From the Canoe River, S.W., over rocky ground to the Divide from the North Thompson, 2,900 feet above the sea	5	..	[?]	Rise 360
Thence down to the North Thompson.. ..	9	..	Fall [?]	450
Thence in a W.S.W. direction over mountainous ground to the Divide from Clearwater River	5	..	[?]	Rise 200
Thence down to the river	7	..	Fall [?]	300
From Clearwater River to the Divide from the Great Quesnelle Lake	3	..	[?]	Rise 150
Thence through a mountainous country S.S.W. to the south-eastern end of the lake [2040 feet above the sea]	10	..	[?]	Fall 460
				..
	55	55	..	
	..	55	..	..

Section.	Miles.	Rail-road. Miles.	Steam Navi-gation. Miles.	Rise. Feet.
Over		55		
2 Thence along Quesnelle Lake to its south-western angle	..	..	45	level
3 From Quesnelle Lake across a slightly rolling fertile country to the mouth of Deep Creek on the Fraser, and below Soda Creek, viz.: From Quesnelle Lake W.S.W. to the Divide, near Round Tent Lake	17	..	..	160
Thence to Deep Creek	10	..	Fall	125
Along Deep Creek west to the Frazer [1450 feet above the sea] with bridge and approaches	8½	..	..	625
				750
Thence W.S.W. across the Chilcoaten Plain to the old Fort on the Chilcoaten River ..	58	..	Rise	697
Thence in the same direction to the mouth of the Gap at the entrance of the Cascade Mountains on the Bute Inlet route	47	..	Rise	200
	140½	140½	..	..
Railroad	..	195½	45	897
Steam Navigation	..	45	..	..
Total Miles	..	240½	..	..
AGAINST				
1 From opposite Tête Jaune Cache to the Rapide des Fourneaux, railroad	10	..	..	..
2 Navigation on the Upper Fraser	280	..	..	..
3 From Quesnellemouth to the Gap, as above ..	137½	427½	..	..
Less distance	..	187	..	..

This road would pass for 40 miles over a wild, unknown, uninhabited, and very mountainous tract of country, between Quesnelle Lake and Canoe River, which would present formidable difficulties and be vastly expensive. Very different from the fertile district on the Fraser and the facilities for immediate navigation.

ALFRED WADDINGTON.

E.—ABSTRACT FROM A PETITION TO THE HOUSE OF COMMONS.

The humble petition and memorial of the undersigned Alfred Waddington, sheweth: &c., &c.,

"That for the purpose (that of an Overland Communication) British Columbia is the key of the Pacific, and that unless a different policy be adopted towards that colony in future, England might be prepared to lose it; owing partly to its distance from the home country and the consequent cost of emigration, partly to its being hemmed in by the United States, but above all to the deep disaffection occasioned by misrule and the arbitrary nature of its institutions, so different from those that surround it.

"That Vancouver Island and the Mainland were till lately separate colonies, with one and the same governor; when unfortunately for both, two distinct governors were appointed, over a total population of ten or twelve thousand souls. That from that moment a system of commercial aggression, if not hostility, towards Vancouver Island was adopted by the Government of the mainland, and things brought to such a state, that in a fit of despair the House of Assembly in Victoria petitioned the Home Government for the union of the two colonies, and, fondly trusting to the liberality of the mother country, offered (without asking the consent of the people) to accept whatever institutions she might think fit to grant.

"That the Governor of the Mainland, who was then in England on leave of absence, was consulted, his views embodied in a bill, and the two colonies shortly after united by Act of Parliament. The representative government of Vancouver Island was abolished without a dissenting voice in either House, the free port of Victoria done away with, and the hostile governor of the Mainland re-appointed over the united colonies, with a Legislative Council consisting of 22 members, of whom fourteen were appointed by himself, and the eight others elected subject to his approval. The capital was removed to New Westminster, a village of 700 inhabitants, the officials there alone maintained, those of Vancouver Island discharged, and the general welfare of the colony thenceforth sacrificed to local interests.

"That although Vancouver Island commands the coasts of the United States on the Pacific as completely as England does those of France on the Atlantic, the Home Government had never spent anything on it; the colony, though yielding over half a million of gold yearly, is indebted to the United States for the carriage of every emigrant and every letter that reaches her shores; and if a man-of-war require repairs she must go to San Francisco; thus depriving the colony of the benefit of an expenditure which ought naturally to accrue to her. In short, the only ties that bind the colony to the mother country are the infliction on it of a despotic form of government; an expenditure over which the people have practically no control, and which is out of all proportion with the means of the colony; and a Governor with a salary of £4,000 a year, besides other allowances, who only consults the interests and wishes of the very smallest portion of the colony; so that although public improvements have been for some time suspended, the colony is sunk in debt, the trade of Victoria has been annihilated, the population has dwindled to a shadow; those who remain are, to say the least, disaffected; and unless some more real interest is evinced for the colony, when the occasion offers she may be driven to vote for annexation to the United States.

"But British Columbia is the key to the North Pacific. Without her and the Saskatchewan territory, the very existence of Canada as a British dependency would be compromised, and before long at an end. The United States are already knocking at the door, and if the whole of British North America is not speedily connected by an overland communication or by railroad, England may bid adieu for ever not only to Canada but to the greater portion of her trade with the East, and, as a consequence, to her commercial supremacy."

"Your petitioner, therefore, humbly but most earnestly prays," &c.

"Tavistock Hotel, Covent Garden,
May 25th, 1868."

"ALFRED WADDINGTON."

F.—CANADIAN STATISTICS.

"A glance at the statistics of Canada will show that her material progress has kept pace with her Legislation. The population of Canada rose from 1,842,265 in 1851, to 2,506,756 in 1861. (It is now with Nova Scotia and New Brunswick four millions). Montreal has a population of 110,000, and Quebec 65,000; in 1831 they each had but 27,000. The population of Toronto has risen since 1842 from 13,000 to 80,000 inhabitants.

"The trade of Canada in exports and imports increased in twelve years, between 1852 and 1864, from 34,342,466 dols. to 91,165,512 dols.

"The number of acres of land held in Canada by private owners in 1852 was 9,825,515, in 1861 it had risen to 13,354,907.

"In 1852 Canada had 3,702,738 acres of land under cultivation; in 1861 she had 6,051,619.

"In 1852 Canada produced 12,802,550 bushels of wheat; in 1861, 24,640,425."

(Letter from the Hon. Charles Tupper to the Earl of Carnarvon, October 19th, 1866).

JUDD AND GLASS, PHŒNIX PRINTING WORKS, DOCTORS' COMMONS.

www.ingramcontent.com/pod-product-compliance
Lightning Source LLC
Chambersburg PA
CBHW031807090426
42739CB00008B/1204
* 9 7 8 3 3 3 7 1 4 8 3 0 0 *